Guide to
Electrical Maintenance

Published by The Institution of Engineering and Technology, London, United Kingdom

The Institution of Engineering and Technology is registered as a Charity in England & Wales (no. 211014) and Scotland (no. SC038698).

The Institution of Engineering and Technology
Michael Faraday House
Six Hills Way, Stevenage
Herts, SG1 2AY, United Kingdom
www.theiet.org

It is the constant aim of the IET to improve the quality of our products and services. We should be grateful if anyone finding an inaccuracy or ambiguity while using this document would inform the IET Standards development team at IETStandardsStaff@theiet.org or The IET, Six Hills Way, Stevenage SG1 2AY, UK.

ISBN 978-1-84919-921-6 (paperback)
ISBN 978-1-84919-922-3 (electronic)

Contents

List of Tables   5

List of Figures   5

Acknowledgements   6

Foreword   7

Author Biography   8

1   Introduction   9
1.1   Objective   9
1.2   The legal requirement   9
1.3   Requirements for electrical installations   11
1.4   Scope of use   11
1.5   Sustainability   14
1.6   Definition   14

2   Aids for good electrical maintenance   15
2.1   Importance of system design   15
2.2   Additional research links   25
2.3   Importance of commissioning and documentation   25
2.4   Correlation of the installation   27
2.5   Maintenance records   33

3   Importance of effective electrical maintenance   35
3.1   Background   35
3.2   Benefits   36
3.3   Consequences   37
3.4   Pattern of activity   38

4   Key steps to electrical maintenance   44
4.1   Background   44
4.2   Key strategies   44
4.3   Maintenance safety   48

5   Evaluation of your electrical system   56
5.1   General overall pattern of maintenance evaluation   56
5.2   Evaluation of electrical systems   58
5.3   Operational security considerations   66
5.4   Monitoring of remedial activities   68
5.5   Auditing of maintenance procedures   68

Appendix A   Checklists   **70**

Appendix B   Suggested maintenance activities   **78**

**Appendix C   Summary of relevant legislation and
                standards**   **125**

Appendix D   References   **126**

Appendix E   Glossary   **128**

Index   **131**

List of Tables

Table 2.1	Maintenance factors to be considered during the design process	16
Table 2.2	Residual risk assessment example	18
Table 2.3	Maintenance factors to be considered when planning space	20
Table 2.4	Maintenance factors to be considered when planning resources	23
Table 2.5	Maintenance factors to be considered when selecting components	24
Table 2.6	Establishing the O&M manual	27
Table 2.7	Example of a labelling schedule	30
Table 2.8	Examples of labels	32
Table 3.1	Benefits of undertaking maintenance	36
Table 3.2	Consequences of undertaking maintenance	37
Table 3.3	Guidance on specific tasks prior to maintenance	39
Table 3.4	Guidance on ongoing tasks to be controlled as part of maintenance	40
Table 3.5	Guidance on tasks after maintenance processes have been established	42
Table 4.1	Example of electrical load operational risk analysis	53
Table 5.1	Types of test and frequency	60
Table 5.2	Electrical installation schedule – test and frequency	61
Table 5.3	Items to be checked and assessed	62
Table 5.4	Guidance on the manufacturer's information	63
Table 5.5	Guidance on equipment failure	64
Table 5.6	Guidance on implementation and failure	66

List of Figures

Figure 1.1	Document structure	13
Figure 2.1	Maintenance resources decision tree	22
Figure 2.2	Schematic diagram with labelling examples	29
Figure 3.1	Typical maintenance process flow	38
Figure 4.1	HSG 85 – Working dead or live	50
Figure 5.1	Plan, Do, Check, Act – the maintenance cycle	57
Figure 5.2	Plan, Do, Check, Act – the developed maintenance cycle	58
Figure 5.3	Failure scenario flow diagram	65

Acknowledgements

IET Standards and the author would like to acknowledge the contributions made by the following individuals to the development of this Guide:

- EUR ING Graham Kenyon CEng MIET TechIOSH
- Simon Robinson CEng FIET FCIBSE MIMechE MSLL
- Bruce Knowles IEng C.WEM FIET MCIWEM
- Russ Hayes
- Martin Hurn MBIFM CDCDP
- Barry Manser EngTech MIHEEM

Foreword

This *Guide to Electrical Maintenance* has been written both as a standalone document and to work alongside other IET standard publications. For example, The IET *Code of Practice for Electrical Safety Management* focused primarily on the health and safety risks of working on electrical installations. This includes highlighting the need for electrical maintenance as an important activity.

This Guide provides guidance on carrying out maintenance activities and using good practice techniques. It examines the operational risks, mitigations and processes that may be used in carrying out electrical maintenance, and also provides insights and philosophies to ensure that electrical maintenance activities are not only safe, but are satisfactorily planned and properly carried out.

The Guide also draws together key guidance from other IET inspection safety and maintenance titles to provide a practical overview for duty holders responsible for maintaining electrical systems, electrical contractors and building service engineers.

This Guide is aimed at electrical technicians, maintenance managers and also at office-based design staff. Good maintenance regimes do not happen by accident: they need careful planning, proactive management and comprehensive reporting. The tone for good maintenance is also established beforehand by considerate design, intelligent construction and satisfactory commissioning.

A good maintenance regime also has its part to play in a more sustainable world where correctly maintained electrical systems keep operating at their maximum energy efficiency and are disposed of correctly at the end of their lifecycle.

Author Biography

Cameron Steel CEng FIET MCIBSE MInstRE
Director, BK Design Associates UK Ltd

Past Chair, IET Built Environment Sector

Cameron Steel began his career in 1982 as an electrician with the Royal Engineers. Following eight years' service he left HM Forces and attended college in Northampton completing a HND Engineering in 1993. He then worked as an electrical contractor in Scotland for 5 years, before moving to building services design consultancy, with RSP in Kent, and working on large Healthcare projects in London.

This was followed by five years with Parsons Brinckerhoff where he was responsible for design and management of building services related projects for the education, healthcare, transportation and defence industries.

In January 2008 he moved to BK Design Associates, an SME M&E Building Services Design Consultancy, where he is now the principal and has been working on healthcare, data centres and government related projects.

Introduction

1.1 Objective

The objective of this Guide is to provide knowledge and guidance on the correct assessment and application of maintenance to electrical systems. The Guide covers various approaches to electrical maintenance to help the reader ensure continuity of service.

Rather than providing definitive answers to every specific electrical installation, this Guide instead provides sufficient guidance to enable an adequate in-house maintenance policy to be written. It will provide links to relevant statutory documents, British Standards and associated publications to allow the reader to undertake a detailed study of his or her own particular installation.

As an initial point of reference the emphasis of this Guide is based primarily on UK-centric documents. The reader should be aware that this is not intended to exclude international or other national standards. Wherever the reader is located it will be possible to find equivalent or similar standards for the electrical systems that are quoted here.

The concepts of good, safe electrical maintenance, and the basic principles that support that, remain the same worldwide and should still be applied following the intent described in this Guide.

1.2 The legal requirement

Within the UK, the Health and Safety at Work etc. Act 1974 places a duty of care on employers, employees and the self-employed to ensure the health, safety and wellbeing of all others at work premises.

There is an obligation on all persons to ensure that the place of work is a safe place. Ignorance or lack of knowledge is deemed to be no defence in law.

The Electricity at Work Regulations 1989 (Statutory Instrument No 635) (EWR) imposes duties on employers, employees and the self-employed to ensure that the safety requirements of the relevant regulations are satisfied. With respect to electrical installations, and the necessity for maintenance activities, Regulation 4 is pertinent:

(1) All systems shall at all times be of such construction as to prevent, so far as is reasonably practicable, danger.

(2) As may be necessary to prevent danger, all systems shall be maintained so as to prevent, so far as is reasonably practicable, such danger.

(3) Every work activity, including operation, use and maintenance of a system and work near a system, shall be carried out in such a manner as not to give rise, so far as is reasonably practicable, to danger.

(4) Any equipment provided under these Regulations for the purpose of protecting persons at work on or near electrical equipment shall be suitable for the use for which it is provided, be maintained in a condition suitable for that use, and be properly used.

In order to fulfil the requirements of EWR it is necessary to ensure that an electrical installation is:

(a) correctly designed and fit for purpose;
(b) correctly installed using the proper materials and equipment;
(c) correctly maintained using appropriate procedures and equipment; and
(d) operated and maintained by competent personnel who have received the correct training.

The Regulatory Reform (Fire Safety) Order 2005 is a statutory document relating to life safety within buildings. Section 13 deals with fire-fighting and fire detection, whilst section 14 deals with emergency routes and exits and also makes reference to emergency lighting. Section 17 (1) states that:

Where necessary in order to safeguard the safety of relevant persons the responsible person must ensure that the premises and any facilities, equipment and devices provided in respect of the premises under this Order ... are subject to a suitable system of maintenance and are maintained in an efficient state, in efficient working order and in good repair.

With respect to electrical installations, the Regulatory Reform (Fire Safety) Order 2005 clearly reiterates the requirement for both fire detection systems and emergency lighting to be installed. It also reinforces the requirement to maintain them so that they are in working order at all times.

The Construction (Design and Management) Regulations 2015 place duties on various stakeholders within the design, construction and commissioning of an installation. In particular, Regulation 9(4) states that:

A designer must take all reasonable steps to provide, with the design, sufficient information about the design, construction or maintenance of the structure, to adequately assist the client, other designers and contractors to comply with their duties under these Regulations.

A badly designed, poorly installed or inadequately maintained electrical installation will place both life and property at risk. Good maintenance on a regular schedule, undertaken by properly trained personnel using proprietary equipment and following correctly managed (documented) procedures, demonstrates clearly that one's obligations under EWR, HASAWA, Regulatory Reform (Fire Safety) Order 2005 and associated documents have been met.

Court cases have been brought, and successful prosecutions obtained, where it can be proved that satisfactory maintenance has not been carried out resulting in accidents or worse, or the potential for the same.

1.3 Requirements for electrical installations

Within the Wiring Regulations, BS 7671:2008+A3:2015, a brief commentary is included in Chapter 34 on maintainability. Regulation 341.1 states that:

An assessment shall be made of the frequency and quality of maintenance the installation can reasonably be expected to receive during its intended life. The person or body responsible for the operation and/or maintenance of the installation shall be consulted. Those characteristics are to be taken into account in applying the requirements of Parts 4 to 7 so that, having regard to the frequency and quality of maintenance expected:

(i)   *any periodic inspection and testing, maintenance and repairs likely to be necessary during the intended life can be readily and safely carried out, and*

(ii)   *the effectiveness of the protective measures for safety during the intended life shall not diminish, and*

(iii)   *the reliability of equipment for proper functioning of the installation is appropriate to the intended life.*

NOTE: *There may be particular statutory requirements relating to maintenance.*

The BS 7671 Sections are covered further: Part 4 discusses 'Protection for Safety'; Part 5 covers 'Selection and Erection of Equipment'; Part 6 deals with 'Inspection and Testing'; Part 7 concludes with 'Special Installations or Locations'.

Section 537.3 of BS 7671 discusses the requirements for switching off for the purposes of mechanical maintenance. Section 537.3.1.2 in particular states that:

Suitable means shall be provided to prevent electrically powered equipment from becoming unintentionally reactivated during mechanical maintenance, unless the means of switching off is continuously under the control of any person performing such maintenance.

Other sections discuss the necessity for any switching device:

(a) to be capable of cutting the full load of the device it supplies (section 537.3.2.1);
(b) to be designed and installed to prevent inadvertent or unintentional switching (section 537.3.2.3); and
(c) to be clearly identifiable and labelled (section 537.3.2.4).

1.4 Scope of use

This Guide builds on the legislative requirements and uses the parameters set out in BS 7671:2008+A3:2015. The Guide aims to assist with prolonging the working life of electrical installations within buildings, and associated areas within the built environment, by providing an understanding of the good working practices for maintaining electrical systems.

The Guide also acts both as a source of reference that can be used by operational estates staff and contractors on site as well as an independent guide, and can be used in conjunction with a number of other IET publications. It should also be referred to by office-based designers to ensure that their work aids the safe working practice of maintenance tasks.

Section 2 of this Guide discusses the importance of engineering design and construction in the implementation of subsequent good practices in electrical maintenance.

Good maintenance regimes are driven by the personnel managing the process, however, it must be recognised that good design provides a level playing field and a strong base for the maintenance to be carried out. Poor design will inhibit this, making life unnecessarily difficult for maintenance personnel and ultimately impacting on costs.

Section 3 discusses the importance of effective electrical maintenance. It describes the actual processes to be used before, during and after maintenance activities and explains the necessity to plan properly and ensure that follow-up activities are not missed. Maintenance of electrical systems can be intrusive and disruptive.

Safe systems of work will often require isolations and interruptions to normal service. Some systems incorporate back up or standby supplies to mitigate such interruptions. However, on reinstatement, it is important that the system is returned either to its original design parameters or to an agreed change in those parameters. This Guide will provide guidance on the safe commencement, as well as safe cessation and satisfactory reinstatement, of maintenance activities.

As well as the hard, or direct, skills required for good maintenance philosophies, the soft, or indirect, skills of design, commissioning and documentation of electrical systems will be discussed.

Section 4 discusses key steps to electrical maintenance and different maintenance strategies. Ideally all activities should be planned; however, systems can fail on occasion, in which cases reactive rather than proactive procedures should be adopted.

In all respects though, whether it be an emergency restoration of a system or a more considered activity, safety and the prevention of danger must be considered the real priority.

Section 5 considers the evaluation of the electrical systems. After maintenance activities have taken place and systems have been restored can the maintenance team actually manage the installation? Conversely, are the processes that are in place (and the budget available) actually conspiring to constrain the team to always be in reactive mode? Can the processes be improved upon so that the team are proactive instead? How do you evaluate an electrical installation in order to try and stay one step ahead of the inevitable decline of the system?

There are tables and flow diagrams shown throughout this Guide that summarise electrical maintenance concepts. The ideas are included not as definitive and all-encompassing criteria but instead as a guide to inform the reader of ideas and concepts around particular themes in electrical maintenance. They can be used as a starting point for further individual research.

Checklists relating to the maintenance stakeholders, and the associated requirements of designs, are included in Appendix A1 and can be used by the reader as a framework to influence design. The format of these checklists is derived from the tables in Section 2. The reader may wish to develop and adapt them to their own particular installation.

Appendix A2 provides further checklists based on the text in Section 2 relating to the activities before, during and after maintenance activities take place. They provide a framework for good maintenance management and, again, can be adapted and developed by the reader.

Within the appendices this Guide will provide a brief overview of the maintenance of several types of electrical systems. However, it will not cover them in depth, nor will it cover specific design issues or criteria. There are many other documents that provide that particular service; where necessary the relevant design documents and publications will be signposted for ease of reference.

▼ **Figure 1.1** Document structure

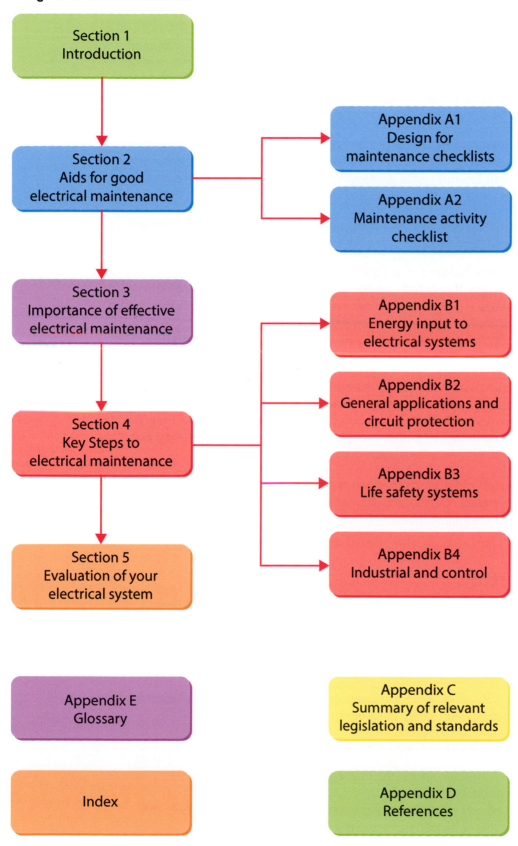

1.5 Sustainability

A side issue of the maintenance of electrical installations, but a very important consideration nonetheless, is sustainability. This can mean different things to different people, but in the context of the built environment, sustainability should mean using less of the earth's resources, whilst maintaining a satisfactory environment for the building occupants at an appropriate level of cost. In other words, sustainability in the built environment closely follows the trinity of people, planet and profit.

Most modern electrical installations are, quite rightly, designed to provide maximum energy savings. For instance, there may well be micro-generation systems (solar panels) on site to offset local energy use. There will be sensors and detectors to control lighting in daylight hours and in times of low occupancy. There will also be meters and sub meters installed to monitor performance and facilitate energy target management.

Comprehensive evaluation processes exist to ensure that the building design complies with legislation such as Part L of the Building Regulations, benchmarks such as Building Research Establishment Environmental Assessment Methodology (BREEAM) and other recognised standards.

Close control is then required throughout the construction and commissioning processes to ensure that the building stays close to the original sustainability concepts.

The performance of new buildings is a topic of some debate within the construction industry, especially when the actual results are compared to the original design parameters. The difference is commonly known as the 'performance gap', which is the gap between the declared design performance and the actual operational performance. This is the subject of continuing research by bodies such as the Chartered Institution of Building Services Engineers (CIBSE), Building Research Establishment (BRE) and others.

Performance gap research may focus on a more realistic appraisal of the design criteria and on the subsequent behaviour of the occupants. However, it is also simple to make the link to a robust maintenance regime as an additional factor in reducing the performance gap. If you do not look after the systems designed to make an installation more sustainable, then the operational performance will not be as efficient.

Satisfactory maintenance of the electrical systems plays an integral part in ensuring that the reality of the day-to-day operational performance of a building matches the design intent and the original theoretical criteria.

1.6 Definition

Within the context of this Guide, 'electrical maintenance' is a series of operational activities designed to prevent danger to people and property. These activities will mitigate the inevitable deterioration of the electrical installation that could be caused by wear and tear, or by adverse environmental conditions.

The purpose of electrical maintenance is to assess and repair or replace, where necessary, and to prolong the integrity and safety of an electrical installation.

Regularly undertaking correctly managed effective maintenance:

(a) will demonstrate compliance with statutory legislation;
(b) should improve reliability;
(c) should maintain energy efficiency and appropriate use of natural resources; and
(d) must reduce the risk of danger and accidents.

SECTION 2

Aids for good electrical maintenance

2.1 Importance of system design

All too often the activities of the design team and the maintenance team are disconnected. For example, the requirement of a new building is identified, it is then designed and constructed and handed over. Unfortunately, it is sometimes only at this point that maintenance regimes are then considered, with the primary focus being on frequency of activities rather than on the practical realities of how an electrical maintenance task is to be carried out.

2.1.1 Relevance of satisfactory design

A more holistic approach would be to fully consider how an installation is to be maintained before the design even leaves the drawing board; a satisfactory design must therefore consider how the installation will be operated. Those considerations need to be carried on through the respective construction and commissioning stages. The legacy of a poor design is all too obvious and the designer must put themselves in the mind-set of those who will operate, maintain and, ultimately, remediate an installation.

Good maintenance does not start by accident or with a project handover. It should start far earlier. Project handover should signal the continuation of maintenance practices that have already been correctly initiated and, to a certain extent, already commenced.

Incorrect design can have significant implications on maintenance resources and costs for several years until it is finally replaced and reconfigured. It should also be noted that over-engineered systems can be just as bad as inadequate designs in this respect, requiring more resources to maintain the installation adequately, potentially wasting energy and more than likely working inefficiently throughout its lifetime.

Within various industries, from rail to construction to manufacturing, there are different stages of design, each with a design and cost review gateway. With each of these design processes the need for adequate consideration of the future requirements of maintenance is quite apparent.

2.1.2 Maintenance stakeholder

Maintenance stakeholders need to be involved in the design process because their experience and local knowledge will inform the designers and assist with design issues.

Likewise there is an onus on the maintenance team to respond to the designers in a timely manner with realistic ideas in terms of budget and practicality of use.

Within table 2.1 some of the likely criteria are discussed. This list may not be exhaustive and consultation on site may reveal further issues that need to be dealt with at the design stage or managed carefully going forward to the operational stage and subsequent maintenance activities.

Operational restrictions	There may be operational restrictions on a particular design solution, perhaps because of clashes with other services or from limits imposed by an inadequacy in the capacity of the existing infrastructure.
Equipment preferences	The maintenance team may have preferences for particular materials or manufacturers and equipment ranges, because of reliability, familiarity or compatibility with the existing installation or neighbouring buildings.
Market availability	Equipment preferences may also be driven by the availability within the local market of replacement components and their associated costs. Are there restrictions, due to in-house procurement policies (such as three known quotes for instance), of specialist equipment from particular suppliers?
Specialist training	Specialist training may be required after handover, which requires budget planning and external courses, for example, training on specific manufacturer equipment or adoption of HV switchgear, which requires Authorised Person certification or outsourcing.
Local standards	There may be local standards that exceed normal industry requirements, for instance, multi-national companies may adopt parent company standards that are more onerous than local requirements.
Derogations	Likewise, there may be long-standing local derogations on a particular standard or working practice where the associated risks are accepted, documented and mitigations are in place.
Downtime	Data centres categorise their availability according to the rule of 9s (90 %, 99 %, 99.9 %, 99.99 % etc.) measured over a year. Hospitals select an infrastructure according to the criticality of the patients being cared for: (a) if a system fails what is the impact of the downtime? What is plan B? (b) are there risks to people's health or safety? (c) is there an impact on commercial processes?

2.1.3 Residual risk assessments

One input to maintenance may be a residual risk assessment left in place following completion of the design and construction of an installation project. Generally the work undertaken, and the electrical building services equipment specified, are commonplace in construction and would be easily recognised by competent contractors. However, where this is not the case the designer should make reference to features and materials that require special consideration.

Suitable site-specific risk assessments should be carried out prior to commencing any maintenance works. The assessment must take into account potential hazards to employees and members of the public that may arise from the works to be carried out.

Residual risk assessments should be carried out by the designers so that site specific details are correctly identified, including evaluation of the potential risks and hazards, together with any measures that need to be taken to minimise or eliminate them. Such details may be based on the original items that were included within the designers risk assessment.

During the design development phase and the construction phase many of those items will have been negated. Those that are left are deemed residual and could impede further design work or subsequent maintenance work.

Such a residual risk assessment should be regularly reviewed and updated by the maintenance team and handed over to a subsequent design team to inform the next development or refurbishment of the same area.

The schedule in table 2.2 below, based on a real example, provides information on some possible hazards and control measures envisaged by the design.

The background to this example is that an area of an estate was being upgraded before the replacement of an existing standby generator, which was a separate project being undertaken by a separate project team. This late replacement of the generator was driven by commercial considerations rather than preferred engineering priorities.

Such occurrences are not uncommon and so it is incumbent on both the design team and the maintenance team to understand the associated risks.

Item	Element	Activity	Design assumptions/ control measures	Notes
1	Electrical infrastructure	Additional circuits	New circuits may be required within the vicinity at some point in the future. The infrastructure of the whole building is being developed. The existing load should be analysed and checked before new loads are added, especially on the maintained sub-distribution board.	Load on maintained sub-distribution board close to fully loaded until a new site generator can be commissioned. Consider providing an additional supply if necessary.
2	Electrical infrastructure	Maintained sub-distribution board power supply	The maintained sub-distribution board is fed from an existing generator panel (100 A TPN). A 160 A supply has been requested for the new generator panel when it is installed.	At the time of project handover the supply to the maintained board is from the existing site generator panel. The supply is planned to be swapped to the new generator panel when that is completed.
3	Electrical infrastructure	Non-maintained switchboard power supply	The main switchboard non-maintained supply has been diverted to the Research Department building interlink.	At time of project handover the supply to the non-maintained board is from the existing site interconnector. The supply is due to be swapped to the new electrical switchboard panel when that is completed.
4	Electrical sub-circuits	Maintained sockets	Socket-outlets with red rockers have been used to highlight circuits supported by maintained power supplies. There will be an interruption in supply whilst the generator starts up in the event of mains power failure.	The circuits have no UPS support as principal equipment already has individual UPS facilities. Any other equipment will need to be similarly supported.
5	Electrical sub-circuits	RCD sockets	In accordance with BS 7671 the socket-outlet circuits have been designed with RCD protection for the use and protection of sockets by all non-trained personnel.	Care should be taken to limit the quantity of equipment with particular types of power supply units to limit the earth leakage current seen in the circuit by the RCD.
6	Mechanical infrastructure	Electrical water heaters	There is no hot water service feed from the ground floor plant. For the second floor hot water heaters have been installed at either end of the accessible loft space.	Replacement heaters will require access to the loft. Both new and old equipment will require man handling through restricted access areas. Additional electrical supplies may be required for any extra water heaters.
7	IT infrastructure	Data cables	The main data comms cabinet for the IT cat 5E infrastructure on the second floor is located with the ground floor switchroom.	Cable lengths of the structured IT wiring is restricted to a maximum of 90 m. Care should be taken on any additional cabling, especially in the offices as lengths to data points here are very close to the maximum.

2.1.4 Space planning

There is always pressure on space when a building is being designed. Plantrooms and risers are often considered dead space by some because they are situated in the back of house areas that cannot be let or occupied by revenue-producing staff. However, without adequate space for the electrical infrastructure and the mechanical plant, the comfort and safety of all occupants will be at risk. Put simply, the building will not function satisfactorily.

Such systems are expensive to procure and will need maintaining to extend their working lives. As such, an important consideration at the design stage is adequate space planning and access arrangements for ease of maintenance.

Section 729 of BS 7671:2008+A3:2015 provides guidance on the setting out of access gangways in larger switch rooms. Similar information can be found in Guidance Note 7: *Special Locations*. In addition, table 2.3 provides details on some other more generic factors that may be considered.

Additionally, regulation 15 of EWR 1989 discusses working space, access and lighting:

For the purposes of enabling injury to be prevented, adequate working space, adequate means of access, and adequate lighting shall be provided at all electrical equipment on which or near which work is being done in circumstances which may give rise to danger.

More commentary on this topic can be found in paragraphs 228-232 of HSE publication HSR25 (*Memorandum of guidance on the Electricity at Work Regulations 1989*).

Design and space planning	Good design will always consider how much space is required for a system in terms of what access arrangements are needed to allow maintenance works to be carried out in an efficient way, for instance: (a) have the ergonomic realities of maintenance personnel standing in front of an electrical panel been thought through? (b) is there enough space to open a panel or to use a screwdriver or spanner? (c) do the riser cupboard doors clash with distribution board covers? (d) have fire alarm interface units been sited at high level on the wrong side of ventilation ducts making access difficult? (e) have fire dampers been located above solid ceilings preventing maintenance access?
Future expansion	Space considerations of future expansion should be considered and noted both in terms of switchboard extensions and capacity of the upstream infrastructure.
Manufacturer's guides	The manufacturer's design guides should be consulted to ensure that there is the optimum amount of space around a piece of equipment. This includes: (a) space from side walls and ceilings to allow for air circulation and heat dissipation; and (b) sufficient space to allow for cover panels to be removed.
Safe access and egress	Consideration of safe access and egress for personnel should be made. For example: (a) does the room require two access points because of fire suppression systems? (b) have other services been coordinated to allow large equipment to be removed and replaced? (c) are ceiling mounted access panels required? (d) has structural steel work been installed for the use of pulley systems? (e) have access hatches through structural floors been installed too?
Structural considerations	Weights of equipment and relative strength of floors to support safe passage of redundant or replacement equipment need to be assessed to ensure that the building is designed to cope with the passage of such equipment, for example, switch panels, large motors, pumps etc.
Manual handling	Consideration of manual handling procedures that do not compromise operator safety should be made for the replacement of electrical installation system components, such as: (a) turning circles; (b) lifting equipment; and (c) ramps etc. This includes door widths from plantrooms and corridor widths.

2.1.5 Maintenance resources

Consideration must also be made at the design stage for the actual personnel and equipment resources that are required so that an adequate maintenance regime is ensured. These resources need to be kept in proportion, both in terms of the critical nature of the particular installation and the likely ongoing maintenance budget.

A complex installation, such as a hospital, will require access to a number of spare components and the systems need to be easy to monitor and maintain. It will utilise in-house staff and outside contractors to maintain an adequate level of services.

On the other hand, a small high street shop within a larger chain may not need quite such a complex infrastructure of supporting maintenance and will typically call in local tradesmen, or perhaps have a business-to-business relationship with an outsourced service company.

It should be recognised that training in-house staff to deal with specialised equipment may not always be cost effective.

The deployment of in-house maintenance staff may quickly be dominated by time spent on these specialised items of equipment, both in terms of regular periodic testing and also maintenance. That, in turn, will divert them away from other more routine day-to-day tasks.

It may be beneficial to employ specialist contractors, or the manufacturer's own teams, as they will be better trained to undertake diagnostic checks, have more experience and can source replacement components more effectively. It will be a balance of costs: whether to use more expensive outsourced expertise against in-house staff, who will be non-productive for other activities.

When decisions need to be made the process used may resemble the decision tree in figure 2.1. Further factors to be considered when discussing skill sets and resources are shown in table 2.4.

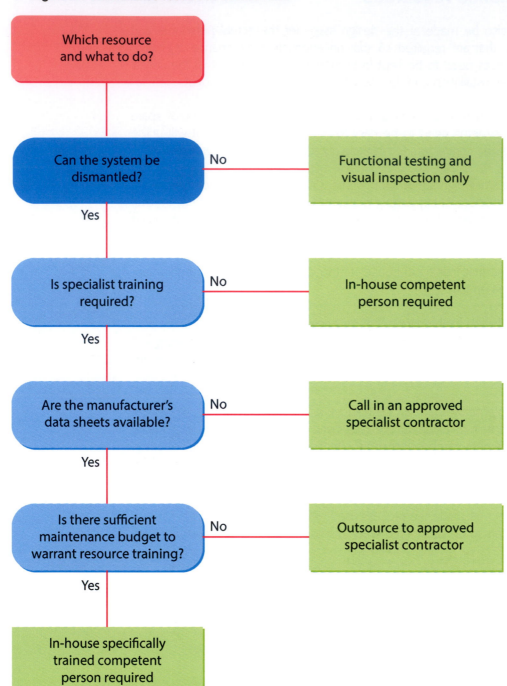

Table 2.4 Maintenance factors to be considered when planning resources

Training at technician level	(a) Is it advisable to have in-house personnel specially trained to maintain particular items of equipment?
	(b) How will their competence level be maintained?
	(c) Is it more expedient to use outside resources?
	(d) Who supervises the process?
Training at managerial level	(a) Is it advisable to have in-house personnel specially trained to manage maintenance?
	(b) How will their competence level be maintained?
	(c) Is it more expedient to use outside resources?
	(d) Who manages the process?
Safe systems of work	Who is responsible for managing the maintenance process when permits and electrical isolations are required?
	The budgetary constraints of training personnel and maintaining competences must be considered.
Manual or automatic monitoring	For example, this may mean the use of an emergency lighting automatic monitoring system instead of local manually operated key switches. Such choices have an impact on both the immediate construction costs and the predicted maintenance budgets.
First line response	The resources for an in-house appropriately experienced maintenance team (complete with associated specialist training update costs) need to be allowed for or, alternatively, outsourced to an approved contractor who can demonstrate the same standards required.
Compliance checks	In-house auditing to ensure appropriate standards are being maintained and that statutory testing of life safety systems is taking place either by senior management or outsourced to an external independent consultant.
Specialist equipment	This is tied into considerations on equipment preferences and specialist training as outlined previously, as is the extra need for specialist equipment to adequately perform maintenance tasks on new electrical equipment.
	(a) Is this already available?
	(b) Does it need purchasing?
	(c) What effect has that on the budget?
	(d) Is it better to simply hire a contractor in on a needs-must basis?

2.1.6 Component selection

Component selection as an aid to maintenance should also be considered. The competing demands of capital expenditure (CapEx) versus operational expenditure (OpEx) are the cause of much disquiet in large estates across many industries.

Ultimately, unless there are healthy budgets for all parties, there has to be compromise and acceptance from both sides that the preferred item of equipment may not always be feasible. Table 2.5 explores in more detail some of the factors to be considered for the correct selection of equipment.

▼ Table 2.5 Maintenance factors to be considered when selecting components

Value engineering	The real definition of value engineering (VE) should be to achieve the same specified outputs at a lesser cost without reducing quality.
	There is always pressure on costs both for construction and for maintenance. However, the introduction of VE to deliver a project may have a detrimental impact on long term maintenance costs, especially if it is misinterpreted and reduces quality as well as cost.
	If there are budgetary constraints there will be value engineering.
	(a) What are the short term effects?
	(b) Can some expenditure be deferred and hence retain quality of components?
Quality	When design selections are made, proper consideration of the quality of components and the declared mean time before failure (MTBF) values need to be considered. This will manage the expectations of the maintenance team in terms of understanding the time before likely failure will happen.
Economy	Cheaper, seemingly more economic, materials and components can cause frequent failures, increased interim resources and hasten end of life replacement.
	Conversely, clever marketing of more expensive components can increase costs when cheaper ones are satisfactory and just as robust.
	BS 8544:2013 *Guide for life cycle costing of maintenance during the in use phases of buildings* will provide guidance on assessing maintenance costs.
Backwards compatibility	Some manufacturers have a policy of backwards compatibility, which can potentially extend the life of some electrical systems. Such a policy allows newer replacement products to be used with legacy systems rather than having to completely replace the existing infrastructure to incorporate newer or recently innovated products.
Obsolescence	Following on from compatibility is the continuing availability of the product, its likely manufacturing life-span and roadmap, and the related key dependencies, including technology progress and migration.
	(a) Will the relentless march of technology advancements make equipment obsolete long before its expected lifecycle ends?
	(b) What impact will this have on maintenance costs?
	To assist with this there are techniques outlined within BS EN 62402:2007 *Obsolescence management. Application guide*.
Product roadmap (system upgrades)	Many electrical systems now involve some form of computerised control. As computer operating systems are updated it may be that some components within an electrical system will no longer be compatible with the operating system.
	For example, can a new software driver be located as a workaround for an old printer when the operating system is changed?
	It is important to emphasise that early adoption of strategies to ensure continuing compatibility and support for all equipment are necessary.
	ITIL ® v3 2011 provides a framework for IT Continuity Management and Change Management.
Stakeholder consultation	Potential conflicts of interest between the capital investment teams and their colleagues in the operational department may cause inadvertent tension and potentially impact on the bottom line of both, not to mention affect the working ethos and politics of the wider company or organisation. Proper consultation of all relevant stakeholders can alleviate this.

2.2 Additional research links

At the time of writing, various industry teams, supported by government departments, are developing new workstream philosophies involving a steady dissemination of knowledge about an installation, from concept design through construction and on to occupation and end of lifecycle of the installation. Known by various names but commonly recognised as building information modelling (BIM), this is set to become a new standard on all government-led contracts.

A UK government paper entitled *Building Information Modelling* (URN 12/1327) defines BIM as *"a collaborative way of working, underpinned by the digital technologies which unlock more efficient methods of designing, creating and maintaining our assets"*.

Whilst many people may perceive BIM as simply a 3D perspective on an installation design that assists with the coordination of multiple services, the whole concept of the BIM project has developed significantly in the last few years. The objective of this initiative now is to provide a coherent and holistic methodology for the design, construction, maintenance and dismantling of the whole building, not just the electrical installation, through all of its lifecycle.

There are a number of resources, accessible online, that explain the various stages of BIM and how good design and the need for ease of maintenance sit within the lifecycle of an installation. More information can be found at http://www.bimtaskgroup.org/.

Additional research, pertinent to the handover between construction and maintenance teams, has also been led by BSRIA under their Soft Landings project to demonstrate the transfer of responsibility from the construction and commissioning teams to the maintenance team. More information can be found at www.softlandings.org.uk.

Previous work in this area by BSRIA includes their building guide document: BG01/2007 *Handover, O&M Manuals, and Project Feedback – A Toolkit for Designers and Contractors.*

2.3 Importance of commissioning and documentation

Assuming that an electrical installation has been correctly designed and installed, the next, and arguably most important, phase in its development is the accurate recording of all commissioning data, schedules and drawings. It is from this point that the operational maintenance of electrical systems begins.

2.3.1 Operation and maintenance manual

Correct information within the operation and maintenance (O&M) manuals is essential to establishing a strong maintenance regime from day one. It should be complete, amongst other documents, with:

(a) accurate drawings;
(b) commissioning data;
(c) test results and certification;
(d) schedules of equipment; and
(e) circuit references.

A number of companies now provide templates for O&M manuals, whilst others actually undertake the work of compiling the volumes of material required. With the advent of BIM and strengthening of the Construction (Design and Management) Regulations (CDM), the standards for building services information in O&M manuals may begin to coalesce. There is a variety of sources available for more information on this topic, for example, http://www.designingbuildings.co.uk with the search term 'Building owner's manual - O&M manual'.

Getting information on specific equipment correct is the responsibility of the manufacturers. BS EN 82079-1:2012 *Preparation of instructions for use. Structuring, content and presentation. General principles and detailed requirements* provides a European harmonised standard on the provision of instructions and assists in defining the content and requirements of O&M Manuals.

The O&M should be treated, not just as a record of the installation at a single point in time, but as a working document (sometimes referred to as a 'live technical file') to which additions and adaptations are noted as amendments. Superseded component documents of the O&M should be archived; this avoids unnecessary confusion by referring to the wrong document.

Continuous updates are crucial to ensuring that the information held within the O&M manual is relevant and accurate. This assists with making maintenance much more straightforward. Logbooks of maintenance activities can then be added to provide a complete record of the lifecycle of the electrical design. Not only will this aid the confidence of the operational team, it could also have a positive impact on the residual value of the electrical installation. An analogy is a second hand car with a complete service record that is, in theory, worth more than a similar car with no maintenance record.

It is also recommended that a logbook is provided as a sign out document for any printed (hard copies) of record drawings. Often such drawings are removed from the O&M manual as a reference document for fault diagnosis within a switchroom or plantroom and then not returned. This is not helpful to ongoing maintenance activities.

In large organisations it is also worth considering a separate server for all electronic (soft) copies of documents. This will enable full access for staff, selected consultants and contractors. However this does give rise to the issue of secure access and cyber security, which is discussed further in Section 5.3.1.

2.3.2 Content of O&M manuals

To set good maintenance practice in motion at the outset, the guidance laid out in table 2.6 should be applied at this stage:

Witnessing of commissioning	On handover of the electrical installation or system, final completion should not be granted until all systems have been witnessed by the client and commissioned satisfactorily by the installation team.
Handover training for operations staff	Handover should include adequate training for all operational staff with particular reference to bespoke switching, such as, a system bypass. This should include all procedures that enable safe systems of work for maintenance purposes.
Site-specific information	Documents need to include site-specific information, not just endless volumes of generic manufacturers' literature and data sheets. This should include a full description of the particular installation.
Settings and default values	The O&M schedules, commissioning sheets and as-built drawings should be complete and provide full information on commissioning settings and default values.
As-built drawings	The chances that design drawings are not altered through the construction process are slim. Final drawings that are submitted with the O&M documents must be accurate and reflect the installation as it finally became commissioned. The contractor must not simply rebadge the original design drawings. The drawings should be carefully updated and amended as necessary before being issued as 'as-built'.
Standard operating procedures	The O&M documents should include standard operating procedures (SOP) to assist in maintenance activity planning and also risk assessments. This will, where appropriate, include switching schedules, workflow sheets and reinstatement routines. An example of this is placing a UPS on system bypass and subsequent placing back into service.
Identification and labelling	Maintenance activities are greatly enhanced by labelling all the principal components in an electrical system. Inadequate identification of cables and systems has serious implications on maintenance. Cables, switchboards, isolators and other equipment throughout a building, for instance, should work to a common labelling system to provide a unique identity for each item. This labelling philosophy should also provide an easy reference to a point of supply.

2.4 Correlation of the installation

The importance of properly labelling electrical equipment with correctly correlated schedules, charts and drawings cannot be overstated as an aid to timely and efficient maintenance. Considerable time and effort, especially in an emergency, can be wasted trying to establish what is connected to what if labelling is incorrect or absent.

Whilst most hands-on maintenance engineers and electricians just want to use their knowledge and get on with the task in hand, proper housekeeping, administration and paperwork are necessary management tools for planning and recording activities.

Regular housekeeping activities should include checks to ensure that existing labels are:

(a) still securely in place;
(b) still legible; and
(c) still relevant.

2.4.1 Labelling

Labels on electrical equipment should ideally be engraved and securely fixed using machine screws or rivets to ensure legibility and robustness. It is possible that local standards may accept typed labels using appropriate handheld devices; however, these may be of more limited use as self-adhesive labels will eventually fall off.

Handwritten labels on insulation tape should only be a temporary solution, perhaps during the construction phase or during the implementation of further additions to an installation, and must be replaced at the earliest opportunity by something more robust and permanent.

Likewise, any handwritten circuit charts must be validated at the earliest opportunity by typewritten test sheets with up-to-date copies left with the relevant distribution board.

The labelling philosophy should ensure consistency throughout the installation and conform to the recommendations of recognised standards including:

(a) BS EN ISO 82079-1;
(b) BS EN61082-1; and
(c) BS EN 60445.

Installations with multiple intakes, perhaps involving more than one HV/LV transformer, use the identity of the various electrical supply origins as the basis of a comprehensive philosophy of the electrical infrastructure labelling.

The example set out in figure 2.2 shows how labelling on a schematic can be developed through the whole installation using the initial supply reference as a consistent prefix. This allows a member of the maintenance team to immediately identify the upstream connections. When transcribed into a schedule, as shown in table 2.7, with further detailed labelling the downstream loads can also be identified.

A series of prefixes to demonstrate the type of equipment, combined with suffixes to determine the infrastructure route, are very beneficial in allowing a circuit to be traced through its means of local isolation, sub-distribution and original source.

Examples of prefixes may be:

SWBD	switchboard	ISOL	isolator
BBC	busbar chamber	DB	distribution board
SW	switchfuse	UPS	uninterruptible power supply
CIR	circuit	MCC	motor control cabinet
ES	essential service (generator back up)	NE	non-essential service

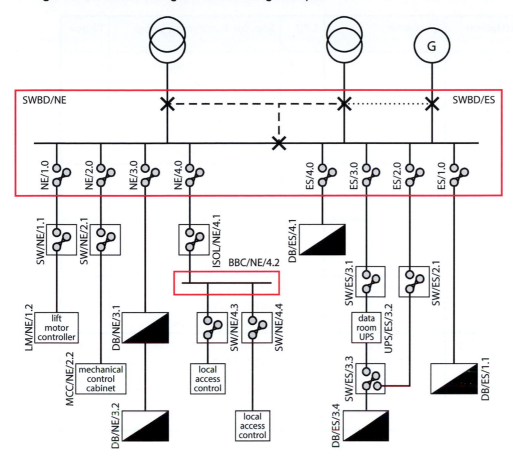

▼ Table 2.7 Example of a labelling schedule

Supplied From	Description	Circuit to	CPD	Text for Equipment Label	Phase
	SWBD/ES/1.0	DB/ES/1.1	100 A	Essential L&P DB supply DBSWBD/ES/1.0 Isolate at main intake	L123
	SWBD/ES/2.0	SW/ES/2.1	63 A	UPS System Bypass SWBD/ES/2.0 Isolate at main intake	L123
	SWBD/ES/3.0	SW/ES/3.1	63 A	UPS System SWBD/ES/3.0 Isolate at main intake	L123
	SWBD/ES/4.0	DB/ES/4.1	80 A	Intake room L&P DB supply SWBD/ES/4.0 Isolate at main intake	L2
	SWBD/NE/1.0	Lift Motor Room	63 A	Lift Motor Room Supply SWBD/NE/1.0 Isolate at main intake	L123
	SWBD/NE/2.0	Mechanical Plant Supplies	100 A	Mechanical Plant Supplies SWBD/NE/2.0 Isolate at main intake	L123
	SWBD/NE/3.0	Local DB for Lighting and Power sub-circuits	100 A	Non-essential L&P DB supply SWBD/NE/3.0 Isolate at main intake	L123
	SWBD/NE/4.0	Access control power supply units	32 A	Access Control power supplies SWBD/NE/4.0 Isolate at main intake	L3
SWBD/ES/1.0	DB/ES/1.1	Essential DB for Lighting and Power sub-circuits	100 A	Essential L&P DB DB/ES/1.1 Isolate at SWBD/ES/1.0	L123
SWBD/ES/2.0	SW/ES/2.1	UPS System bypass	63 A	UPS System Bypass SW/ES/2.1 Isolate at SWBD/ES/2.0	L123
SWBD/ES/3.0	SW/ES/3.1	UPS System	63 A	UPS System SW/ES/3.1 Isolate at SWBD/ES/3.0	L123
SWBD/ES/4.0	DB/ES/4.1	Intake Room DB Lighting and Power sub-circuits	63 A	Intake Room L&P DB DB/ES/4.1 Isolate at SWBD/ES/4.0	L2
SWBD/NE/1.0	SW/NE/1.1	Lift Motor Room	63 A	Lift Motor Room SW/NE/1.1 Isolate at SWBD/NE/1.0	L123
SWBD/NE/2.0	SW/NE/2.1	Mechanical Plant Control cabinet	100 A	Mechanical Plant Supply SW/NE/2.1 Isolate at SWBD/NE/2.0	L123
SWBD/NE/3.0	DB/NE/3.1	Local DB for Lighting and Power sub-circuits	100 A	Non-essential L&P DB DB/NE/3.1 Isolate at SWBD/NE/3.0	L123
SWBD/NE/4.0	ISOL/NE/4.1	Isolator for access control power supply units	32 A	Access Control Supply isolator ISOL/NE/4.1 Isolate at SWBD/NE/4.0	L3

Supplied From	Description	Circuit to	CPD	Text for Equipment Label	Phase
SW/ES/3.1 & UPS/ES/3.2 or SW/ES/2.1	SW/ES/3.3	Data Room DB	63 A	UPS System SW/ES/3.1 Isolate at SWBD/ES/2.0	L123
SW/NE/1.1	LM/NE/1.2	Lift Motor Room	63 A	Lift Motor Controller LM/NE/1.1 Isolate at SWNE/1.1	L123
SW/NE/2.1	MCC/NE/2.2	Mechanical Plant Control cabinet	100 A	Mechanical Plant Control cabinet MCC/NE/2.2 Isolate at SW/NE/2.1	L123
DB/NE/3.1	DB/NE/3.2	Sub-distribution board for Lighting and Power sub-circuits	63 A	Non-essential L&P DB DB/NE/3.2 Isolate at DB/NE/3.1	L123
ISOL/NE 4.1	BBC/NE 4.2	Busbar Chamber for multiple power supplies	100 A	Access Control Busbar Chamber BBC/NE/4.2 Isolate at ISOL/NE/4.1	L3
BBC/NE 4.2	SW/NE 4.3	Switched fuse outlet for access control power supply unit	5 A	Access Control Switched Fuse SW/NE/4.3 Isolate at ISOL/NE/4.1	L3
BBC/NE 4.2	SW/NE 4.4	Switched fuse outlet for access control power supply unit	5 A	Access Control Supply isolator SW/NE/4.4 Isolate at ISOL/NE/4.1	L3

Examples of good labelling practice	

Examples of poor labelling practice	

2.4.2 Standard operation procedures

Standard operation procedures (SOP) and workflow sheets are pre-prepared documents that can greatly assist with normal maintenance tasks, especially for personnel who are relatively new to the site. Accurate descriptions are vital with clear, concise and unambiguous statements. Aided by prepared drawings where necessary (such as diagnostic fault trees and pictorial diagrams), these documents will act as important aids to reduce mistakes.

Such documents, together with prepared switching schedules, can also aid method statements for general supply changeovers such as uninterruptible power supply (UPS) swap outs. For other reactive maintenance tasks or adaptations to the existing system, specially prepared documents may be required to demonstrate a proper understanding of the task.

Many electrical systems have key operated mechanical interlocks to allow for the connection of multiple supply points for resilience. The interlocks are marshalled using key exchange boxes and provided to inhibit the application of more than one point of supply at any one particular time. Prepared operation procedures allow operatives to be guided through the process of changing supplies.

Examples of exchange boxes for key operated mechanical interlocks

These SOP documents should be located in the site O&M manual. Some sites also provide copies of the SOP documents within the switchrooms themselves for ease of reference.

Other installations, with high levels of security, may make the local decision not to provide copies of these documents within the switch room, even though such installations will have quite restrictive access arrangements.

SOPs should provide clear, precise instructions with diagrams or annotated photographs. For example, with interlock switching on main switchboards, the instructions should clearly explain how to change to alternative supplies and how to subsequently reinstate back to the original status. Copies of check sheets on switching arrangements could also be attached to any logbook to demonstrate what has taken place.

Workflow diagrams can either supplement the SOP or act as a standalone pictorial sheet to guide the electrical technician or site engineer through the maintenance task that has been set.

It is also important that SOPs are regularly reviewed to ensure that they are relevant and up to date. This review should take into account any equipment changes or alterations to the electrical infrastructure. Relying on an out-of-date SOP, especially under emergency conditions, could represent a safety hazard. Section 5.5 of this Guide discusses the auditing of maintenance procedures and reviewing the SOPs could form part of that process.

2.5 Maintenance records

2.5.1 Logbooks

Close administrative controls on maintenance should always be given; this should ensure a safe system of work and a handy reference for similar future activities too. However, on occasion, things can, and do, go wrong. A proper record of the activities that took place will assist in establishing what mistakes were made (if any) and how to avoid similar occurrences again. A comprehensive record of electrical maintenance activities can also assist in establishing that a component failure or process mistake was outside of anyone's control.

Maintenance logbooks are a means of carrying this out and, if used correctly, can be a useful tool to collate knowledge of the systems. Paperwork can be perceived as an administrative burden, especially to those that are more practically minded. However, if viewed in a constructive manner, it is actually a concise log of activities, preferably recorded in real time, which can greatly assist all of the maintenance team going forwards.

Additionally, within any organisation, there is inevitably a turnover of personnel. As each departure happens there is an inevitable loss of local knowledge of the electrical systems. Providing a comprehensive log of previous activities will greatly assist new personnel to become familiar more quickly with the particular issues in that electrical installation. As such, an up-to-date site logbook can also provide a cost effective element of an on-site training programme for new staff.

On larger installations a safe system of work will involve permits to work and associated documents. These will also be cross-referenced in a logbook to provide a chronological record of activities.

2.5.2 Life safety systems

As discussed earlier in Section 1.2 with the Regulatory Reform (Fire Safety) Order, it is an important statutory requirement, within public buildings, that life safety systems are regularly tested in accordance with the timetables laid out in the relevant British Standards, for example, BS 5839 for fire alarms and BS 5266 for emergency lighting.

A written record of these tests within a logbook, ideally kept in a central control location, is important for maintaining a history of these tests. When testing systems are automated, printouts of the timings of the tests and associated results should be retained for future analysis. Whether by hand or automatic printout, this all provides evidence for audit purposes, be that internally from an authorising engineer or an external organisation, such as the local Fire and Rescue Service.

The requirements of these particular British Standards are often such that a sample of installation points across the whole premises, for example, break glasses for fire alarms in one zone or emergency luminaires in one particular escape corridor, are tested on each occasion and that this is varied at each subsequent periodic test. If tabulated correctly a logbook serves as an ongoing record of which areas have been tested previously and over the course of a year it can be demonstrated that all parts of the installation have been tested satisfactorily.

Importance of effective electrical maintenance

3.1 Background

An electrical installation has been completed, commissioned and set to work. It operates satisfactorily and maintenance is an overhead – so if it works why bother to do maintenance at all? It's under warranty after all, isn't it? Well, yes, but not if it is misused or not maintained in accordance with the manufacturer's recommendations. In that case the warranty may well be revoked by the manufacturer.

Electrical systems, by definition, involve the conversion of electrical energy from one type of energy to another. Each stage of conversion has varying degrees of efficiency with heat loss as a by-product. Over time this heat loss has a detrimental effect at component level, hastening the end of the lifecycle for the component and subsequently causing system failure.

As an added complication, an electrical system may be installed in a harsh environment that degrades its performance. Examples of this would be the illumination levels within a factory or the integrity of electrical equipment IP ratings at a vehicle wash down point. The environment in which an electrical system is installed may also change rapidly, for instance if the building infrastructure fails and a leaking roof leads to catastrophic failure of electrical equipment.

Electrical systems are, hopefully, designed for optimal performance. Clearly though, if a system degrades over time, that optimal performance will be lost. If an electrical system is cheap to replace it might be considered that maintenance is not necessary. However, the reality is that, more often than not, electrical systems are expensive to replace at the end of their lifecycle.

The application of a robust maintenance regime will extend the life of an electrical system, occasionally beyond the manufacturer's sales pitch and the original design lifecycle criteria.

To reinforce this, certain electrical systems, typically related to life safety functions such as fire alarms and emergency lighting, have periodic maintenance and functional testing regimes imposed on them by the requirements of statutory documents and associated European and UK standards. These ensure certain minimum operational standards are upheld so that these systems will operate satisfactorily when required to do so.

Within the following sections, tables 3.1 and 3.2 explore the various benefits and consequences of undertaking electrical maintenance.

3.2 Benefits

▼ **Table 3.1** Benefits of undertaking maintenance

Extends lifecycle	It assists with extending the longevity of an electrical system in most types of installation environment.
	Whilst it must be accepted that there may be component level failures and replacements, such as lamps or fuses, the equipment system will continue operating for longer periods following some basic remedial activity.
Improves performance	It provides an ongoing assessment of the performance of the electrical installation to ensure that it is at, or near, the designed performance parameters.
	For instance, cleaning dust, insect debris and cobwebs from luminaires to improve light levels, or in checking neutral cable connections in plantroom installations with high harmonic loads such as variable speed drives.
Less downtime	It may mean fewer unplanned interruptions, especially on systems installed in less harsh environments. Systems installed in harsh environments, for instance externally or in factory paint shops, will fail more often.
	However, other areas, with careful planning and regular maintenance, will operate more reliably and will be less likely to fail. Some industries, such as data centres, have quite stringent requirements to minimise downtime.
Manufacturer's warranties	It also adheres to some manufacturers' long-term warranty requirements. Some extended warranties are not valid unless the operator can demonstrate that the equipment or system has been maintained by an approved maintainer in an agreed way during a set period immediately after commissioning. This period will typically be over and above the original twelve months that is the industry norm.
User confidence	It provides reassurance and confidence to users that the systems will operate as required when needed.
	This is especially important with life safety systems such as fire alarms and emergency lighting that will typically be ignored on a day-to-day basis but will still be expected to work satisfactorily in emergency situations such as a building evacuation.
Statutory requirements	It fulfils statutory requirements, in terms of the EWR, HASAWA, Regulatory Reform (Fire Safety) Order 2005.
	It also fulfils the statutory requirements of the enhanced operational needs of emergency lighting and fire alarms, to ensure that such life safety systems have the proper level of output and that they will work reliably when required given that they are typically dormant in most other scenarios.

3.3 Consequences

▼ **Table 3.2** Consequences of undertaking maintenance

Temporary removal of services	The simple act of maintenance may involve taking an electrical system temporarily out of service. This needs planning, resources, replacement components, notification to users, some element of inconvenience and contingency. Critical areas will need alternative power supplies.
Appropriate level of maintenance	It is for a suitably qualified operator to determine if an electrical system should be maintained. Part of that assessment is an analysis of: (a) the associated dangers of not carrying out maintenance; (b) what happens if the system fails; and (c) the potential liabilities that result.
Control and management	The act of carrying out maintenance without adequate controls or planning, i.e. badly, is equally wrong. Without controls a system could catastrophically fail; without planned resources and replacement components the system could be out of action for an extended period of time.
Correct reinstatement	As previously mentioned, maintenance is a necessary part of prolonging the lifecycle of an electrical system. Typically, though, maintenance is intrusive. This carries risk, therefore, care should be taken to ensure that where electrical equipment is isolated or partially dismantled for maintenance purposes it is set back into service correctly. Remember that reinstatement of a system is effectively a form of recommissioning.
Increasing level of resources	Maintenance requires resources, in terms of both operatives and of spare parts, and as the installation ages more may be required to provide equitable levels of service. Budgets and contingency planning for these resources must form an important part of any maintenance management regime.
Product obsolescence	A comprehensive O&M manual will form part of the commissioning package of any electrical installation. Equipment manufacturers regularly develop new products and older ones then become obsolete. (a) It should be recognised that the business is in control of the obsolescence management strategy. Responsibility for that strategy may be delegated to the maintenance team. (b) Decisions will need to be made on whether to buy in spare parts while they are still available or replace with a similar product when the original fails. (c) It should also be noted that circumstances may change due to reasons beyond anyone's control; for example, a business failure in the supply chain or overall system technology obsolescence. In these instances a re-evaluation of the obsolescence strategy will be necessary. It may be prudent to consider a regular review in line with BS EN 62402:2007 *Obsolescence management. Application guide.*

3.4 Pattern of activity

The general pattern of any electrical maintenance regime is largely the same, with activities before, during and after the maintenance event. Good planning and comprehensive reinstatement routines are as important as the central hands-on maintenance activity itself. Adequate resources and time must be allowed for in all three main stages.

It should be remembered that maintenance regimes are not always one-off linear activities, but can be seen as part of a cyclical process with repeat activities throughout the lifecycle of the installation. As the installation ages it is possible that each activity within the cycle will require more manpower planning and material resources. This will be explored further in section 5 of this Guide.

Figure 3.1 demonstrates a typical maintenance process flow. The process flow adopts a typical traffic light colour sequence:

(a) red for planning and preparation;
(b) amber for set-up activities and the actual maintenance task; and
(c) green for system checks and set back into service.

▼ **Figure 3.1** Typical maintenance process flow

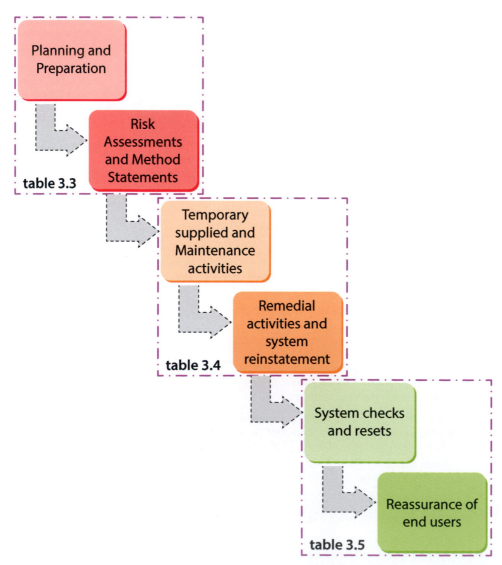

Sections 3.4.1, 3.4.2 and 3.4.3 have been based on an example of a generic switchboard maintenance activity. The scenario allows certain criteria to be examined and so forms the action plan going forward.

Additional criteria should be considered by the reader depending on the stage of the process and the particular type of electrical installation that is being maintained.

In many larger installations, such as hospitals, data centres and larger factories, switchboards are installed that may have multiple intakes, bus-sections and multiple outputs to loads, for example UPS with system bypass.

3.4.1 Before the event

▼ **Table 3.3** Guidance on specific tasks prior to maintenance

Criteria	Examples and explanations
Training and familiarisation	• It is likely that some of the equipment will require specialist technician level training to maintain it correctly. • Regular reviews of the skill sets of the maintenance team must be undertaken to ensure that there is adequate coverage of all specialist equipment. • Regular toolbox talks for first line repairs and responses helps keep knowledge up to date and, if organised properly, perhaps with manufacturers, can add to an individual's continuing professional development.
Outsourcing	• Where specialist technician level skills are necessary a commercial decision may be that the role must be outsourced. This could be the case for annual activities where it is not economic to retain underutilised, highly trained staff. • Solutions here could range from a simple list of preferred local contractors to formal contracts that detail scope and response times.
Contingency planning	• Detailed consideration of what activity is actually planned must be made. From that, risk assessments should be used to analyse what additional contingency should be in place if a maintenance activity cannot be completed, for instance, if a new part is required from a particular manufacturer on a bank holiday weekend.
Resources	• Does the activity itself require additional personnel either with specialist training or with more general skills? • For instance, a major overhaul of a buildings automatic fire detection system will require interruptions to that service. Regular out-of-hours patrols by fire wardens can mitigate this until the fire alarm system is recommissioned and returned to service.
Risk assessments and method statements (RAMS)	• As separate documents, both the risk assessments and the method statements (collectively known as RAMS) should be prepared well in advance of any activity by the person who will be directly responsible for the maintenance activity and checked by all principal stakeholders for comment and for information purposes.
Permits to work, safe systems of work and logbooks	• In conjunction with the RAMS, these documents should be prepared to ensure safe systems of work, including switching schedules, isolation of electrical systems, locking off of switches and other appropriate isolation points. Consideration of particular electrical safety warning notices, declared lock off points and associated padlocks should be made at this time too.

Criteria	Examples and explanations
Notifications and consultation of end users	• Any maintenance on any electrical system, which requires it to be taken temporarily out of service, must be discussed with the end users in that area. • For instance, in some areas of some hospitals there are refrigerators that must not be switched off for any length of time. More importantly, there will be critical patient areas with permanent feeds required to clinical equipment. Alternative arrangements need to be made in these circumstances.
Notification to other occupants	• Any public areas in larger installations must have warning and hazard notices placed to alert all users to the maintenance activity if deemed necessary. • For instance, when changing luminaires in corridors, alternative routes must be sign posted and work areas cordoned off.

3.4.2 During the event

▼ Table 3.4 Guidance on ongoing tasks to be controlled as part of maintenance

Criteria	Examples and explanations
Task briefing and management	• Immediately before a maintenance activity a full briefing of all personnel should be made; this includes any RAMS. • Good practice in most industries is for these documents to be countersigned by all the operatives to acknowledge their understanding of the contents of the RAMS.
Standby supplies, bypass supplies	• Where isolations are required for maintenance activities some installations may provide for alternative electrical supplies. These will probably be switched over using a permit to work (PTW) system and also a mimic diagram update. • Adequate provision for generator fuel should be made where these are used. It is also prudent to use an experienced operator to monitor a generator and check the engine parameters during operation.
Temporary supplies	• Temporary supplies may also be introduced into an area. Where these comprise of extension leads for the local use of safety low voltage hand tools there is a duty of care to ensure that the leads are operating at an appropriate safety voltage and protected by an adequate automatic electrical disconnection system (RCD or similar). • The cables should also be mechanically protected to avoid unnecessary damage and routed to avoid pedestrian and trolley traffic.

Criteria	Examples and explanations
Electrical isolations	• Electrical safety management criteria must be followed carefully. The default policy should be that no live working should take place. • An installation needs to be proved dead before any works take place in accordance with industry working practices. • Where working live can be fully justified (in terms of the requirements of Regulation 14 of the Electricity at Work Regulations), then only appropriately trained staff using the correct personal protective equipment (PPE) and tools should closely follow agreed methods.
Maintenance	• Carrying out actual equipment isolation and the associated maintenance activity in accordance with the previously agreed RAMS documents and scheduled tests and inspection criteria.
Management	• For lengthy tasks it may be deemed prudent to provide regular checks and reviews from a site supervisor or engineering manager to ensure that the task is still being performed in accordance with the previously agreed safety criteria. • If an activity is delayed, or if the project encounters unexpected problems, it is all too easy to drop discipline to maintain previously programmed deadlines. Instead, contingency plans should be initiated and key stakeholders informed. • These plans should also involve monitoring of adverse working conditions, maximum time limits on working in such conditions and appropriately trained relief personnel.
Remedial (system critical & urgent)	• Where possible any remedial activities should be conducted during the maintenance period, especially if isolations have been carried out, where materials are available. • If not then the details of required components should be noted for future reference and a further shutdown. • Consideration of remedial activities along the lines of the four categories in the BS 7671:2008+A3:2015 'Test and Inspection' schedules should be made to prioritise remedial works.
Reinstatement	• Following completion of a maintenance activity the installation needs to be reinstated to its normal running mode. This effectively means an abbreviated commissioning process to ensure that all switches and isolators are in the preferred operational mode and that any indicators, mimic diagrams and other monitoring systems are restored. • Pre-prepared checklists based on local experience and knowledge of the systems could assist here.

3.4.3 After the event

▼ Table 3.5 Guidance on tasks after maintenance processes have been established

Criteria	Examples and explanations
Housekeeping	• Where necessary basic housekeeping tasks need to be done, with all areas checked and cleared of rubbish and debris. • Electrical switchrooms and service risers should not be used as stores or dumping grounds. They inhibit normal operations and maintenance and present a fire risk. • Any unused materials should be returned to a storage area for reuse on other parts of the estate as appropriate. • All tools should be accounted for and removed from the work area.
Administration, permits etc.	• Any administrative paperwork relating to the task must be completed immediately after the conclusion of the maintenance activity, whilst that activity is still fresh in the mind. • Where reinstatement checklists are used these need to demonstrate that an electrical maintenance activity has been completed and the system has been satisfactorily set back into service. • All permits need to be cancelled and filed correctly for future reference or auditing purposes.
Operator debrief/ activity wash up	• It is useful to set aside time to debrief maintenance staff and to check for any issues that may have arisen during the completion of the task. • Any known technical issues can be discussed and added to the ongoing remedial list. • Matters of immediate concern can be escalated if required.
Reassurance of end users	• It is important to ensure that all systems are running in a manner that it is satisfactory to local users. • Any new user interfaces need to be clear and additional guidance should be provided to end users where necessary. • Within a busy hospital, for instance, nursing staff need to know that their own controls work intuitively. They have more pressing work to do than supplementing the work of the engineering staff.

Criteria	Examples and explanations
Follow-up checks	• Where a system has been taken temporarily out of service to perform a maintenance task and subsequently reinstated, checks should be made after an appropriate period to ensure that it is working satisfactorily. • This cursory check should be over and above the reassurance of end users and could be as much as 24 hours later.
Review	• For larger maintenance tasks, or activities involving a life safety system, it may be useful to double check all associated paperwork and documents within 3 working days to ensure that there are no gaps in the records. • At some point, in larger installations, there may well be an audit undertaken. This may not happen for several months – any gaps in the records will be noticed at this point and will need to be accounted for. This could prove difficult after so much time has elapsed.
Remedial (non-critical and less urgent)	• Remedial works deemed as a lower priority should be managed to ensure completion. • Maintenance backlog lists may be used for this. Items that cannot be completed within an agreed period could be placed on the site risk register. This strategy would work for systems that have been earmarked for replacement in a future financial period and where spending more maintenance budget in the interim could be thought as not viable.

Key steps to electrical maintenance

4.1 Background

There are several key strategies that will ensure that the effective maintenance of electrical systems is conducted. Each strategy should be implemented within a framework of a safe system of work. The strategy to use will depend largely on whether the work is carried out before or after failure of an electrical system or its components. Some strategies will seek to prevent failure; others will be more reactive to correcting faults after they have occurred.

Each electrical system needs to be assessed as to its importance to an installation so that it is clear which is the best strategy to use.

For safety systems, such as emergency lighting or fire alarms, a proactive approach is necessary to ensure that the risk of faults is mitigated and the systems operate satisfactorily at all times, especially in periods of maximum occupancy.

For localised systems, such as side office lighting, a more reactive approach, whereby attention is only given once the system has failed, may be more appropriate.

4.2 Key strategies

Within BS 6423:2014 *Code of Practice for the maintenance of Low Voltage Switchgear*, paragraphs 3.8 to 3.11 cover four principal sub-definitions of maintenance:

(a) preventive maintenance (also known as preventative maintenance);
(b) corrective maintenance;
(c) post-fault maintenance; and
(d) predictive maintenance.

Essentially though, this Guide will focus on three principal approaches to maintenance strategies for electrical systems:

(a) Preventative
This simply replaces key components before they completely fail so that the whole system maintains an output that is close to the original design.

A simplified example of this is changing all lamps within school hall luminaires regardless of whether they are working or not when a predefined period of time has elapsed, which may also coincide with a convenient holiday period that allows access.

Another example is replacing the lubricating oil of a standby electrical generator after a certain period of time, irrespective of the number of hours it has run.

(b) Reactive

Similar to the corrective and post-fault maintenance sub-definitions highlighted in (b) and (c) above, this means waiting for a component to fail and then instigating remedial activities to restore the system to its previous operational level.

An obvious example of corrective maintenance is waiting for a lamp to fail under normal operation and then replacing it, whilst ignoring the performance of the lamps within neighbouring luminaires.

An example of post-fault maintenance is replacing a surge protective device on a low voltage switchboard after it has failed safely following a lightning strike on the building.

Both approaches are similar in that they react to events.

(c) Predictive

This more proactive approach monitors the output of a system before it completely fails. Methods based on statistical analysis can take place and as certain predefined parameters are reached then pre-planned remedial activity takes place.

A simplified example of this is changing all lamps within school hall luminaires once a percentage of them have failed, which may also coincide with a convenient holiday period that allows access.

Another example would be regular checks on the lubricating oil of a standby electrical generator, to assess whether the oil is actually ready for a change rather than, as per preventative maintenance, waiting for a certain period of time to lapse.

4.2.1 Preventative maintenance

Under this regime the management philosophy is relatively simple. The electrical systems are catalogued and a simple schedule, to be based either on the rudimentary information left in the contractor's O&M manual or in the manufacturer's recommendations, guides the operators as to when a system should be overhauled and maintained. It is simply based on hours run or calendar-driven dates (for instance, commissioning date anniversaries).

It is important to remember that preventative maintenance does not prevent system- or component-level failure. It simply delays inevitable failure and seeks to restore a system back towards its initial commissioning parameters following a period of deterioration through fair wear and tear. Such an approach can also be over-prescriptive and means that system components are occasionally replaced when some useful output still remains. There are inevitable cost implications to the maintenance budget in terms of materials and resources, with the frequency of maintenance often being higher than the optimum.

The additional level of waste material that is caused by this methodology may be viewed as an issue if the company owning the asset has set its sights on a rigorous sustainability agenda. Another viewpoint is that some organisations may see electrical maintenance as an overhead that should be reduced as much as possible.

However, for some installations, such as factories and schools, there might be some advantages in this policy as the remedial activities are often only permitted to take place during quiet extended holiday periods, when there are fewer people about and because minimum standards of output from the systems are required for safety at all other times.

Another form of preventative maintenance is periodic inspection and testing. For a building's electrical wiring installation reference should be made to BS 7671:2008+A3:2015 and the associated Guidance Notes published by the IET. Within Guidance Note 3 there are defined periods for testing and inspecting an installation. In theory, this should identify potential faults so that planned, rather than reactive, remedial activity can take place.

4.2.2 Reactive maintenance

The management philosophy for this regime is simpler – if it works, why fix it? Wait for it to break first.

This approach may be appropriate for some parts of an electrical installation that are not in critical areas where there are no life, property or operational risks. However, if a system fails and requires fixing there is inevitably a time lag before it is finally working again. Often two visits are required – one to assess the problem, then departure to gather the components required before a second visit to fix and restore the system.

There are some cost advantages in that fewer materials are used because replacement only occurs on complete failure of the component. However, there could be cost disadvantages too if the component failure in turn isolates an entire system because of a single point of failure. Another disadvantage is that any failure could involve more work, as a catastrophic system failure could involve more than one component or sub-system in a domino effect.

A complete risk analysis of all electrical systems should be taken to ensure that all critical systems have action plans for replacement components in the event of failure. Questions to ask include:

(a) how critical is the system?
(b) will the factory production grind to a halt?
(c) is a life at risk in a hospital operating theatre due to an engineering component failure?
(d) is a laissez-faire approach to maintenance in this particular instance good enough?

Time for remedial activities can be reduced if the component selection at the design stage has been thought through and leading manufacturers have been specified with off-the-shelf replacement parts. There can be a tendency at the design stage to select components and systems with recognised performance levels. Again, this involves capital cost and may stifle innovation to a certain extent.

Another strategy could be to hold a stock of critical spare items locally within a secure store room. This does involve capital cost for parts that are rarely used and largely gathering dust, but in the early hours of a weekend morning, for example, restoring a data centre switchboard with parts held in an on-site store room is preferable to waiting for the wholesaler or manufacturer to open on Monday.

Such an approach may work for a private sector client, such as a commercial data centre, where a financial overhead of spare parts lying dormant on a shelf can be accepted. However, the same philosophy may not be so easy at a hospital where the budget must include both the retention of spare parts and the staff budgets. In such a situation there will be a balance as to what can be achieved.

There is another viewpoint with specific regards to legacy systems, which possibly needs to be seen within the context of the wider estate – is it actually better to let a legacy system fail? Such a 'waiting to fail strategy' may be preferable in certain circumstances. It postpones the point at which new investment is required until the optimum moment.

Within this context, certain decisions need to be considered, such as:

(a) are there newer components available that can be introduced to the legacy
system and so extend its life?
*Some manufacturers do provide backward compatibility as they develop their
product offerings.*

(b) is complete replacement more cost effective?
*A cost-benefit analysis with subsequent reviews may determine the best time to
actually replace the component rather than deferring and making do.*

(c) what part does the system play within the overall infrastructure?
*A less critical system that will not bring a larger installation to a standstill may
be allowed to fail. A more critical system needs closer monitoring and potential
replacement sooner rather than later.*

(d) what overall effect will failure have?
*Catastrophic failure of an electrical system without an adequate recovery plan
already in place will obviously undermine the installation and be much harder to
reinstate.*

4.2.3 Predictive maintenance and monitoring

The management philosophy for this regime is more complex as it is based on predicting
when the maintenance service is required, by means of regular checks to look for agreed
levels of deterioration, and then intervention to fix it.

This approach involves measuring certain parameters on the performance of a system
on a regular basis and, when predetermined criteria are reached, intervening at that
particular point to carry out the maintenance to restore the system to its designed levels
of performance.

At its simplest it involves a maintenance operator working around the site with a clipboard
and pen noting the current status of equipment against its intended purpose. In the past
this would have been a paperwork-led exercise with the outputs then analysed by office
staff so that future works could be planned.

The advent of better technology means that often this approach can be recorded by
sensors and collated on a computerised system for analysis by a technician or engineer.
Such a system involves a lot of capital investment in terms of technology and personnel.
In reality it will only be used on more expensive plants or equipment in critical areas.

In recent years the advances in Building Management Systems (BMS) and IT cabling
infrastructure mean that technological assistance can be given to maintenance monitoring.
A number of electrical systems that previously worked completely independently can
now be linked back to a common user interface where common faults can be flagged
up automatically.

Developments in Power over Ethernet (PoE) and also the Internet of Things (IoT) provide
more channels for automated monitoring for maintenance purposes.

4.2.4 Considerations for computer-aided maintenance monitoring

A number of companies now produce building-wide systems that provide a high degree of intelligence for integration of various building infrastructures. This provides a number of advantages for operational purposes in terms of saving energy, and for maintenance purposes in terms of measuring the performance of electrical systems, planning associated remedial tasks and recording activities to close out any outstanding issues.

Clearly, such systems feed into BIM from the operational perspective and a suite of new standards are being developed under the banner of PAS 1192.

Increasingly, these computer-aided systems are web-based and so can be accessed externally by key personnel. There are also some considerations to be made with such systems; there is increasing awareness of the implications of cyber security. Initiatives by the IET, including Codes of Practice, and by other organisations provide guidance on using this methodology to manage systems remotely whilst remaining safe from external, unauthorised interference.

Additionally, there are sophisticated methods for implementing the asset management of both mechanical and electrical systems. Some facets of asset management will naturally feed into maintenance monitoring, which involves gathering data and statistical analysis to inform maintenance strategies. This is becoming a field of expertise and some academic institutions are now running Masters-level degree courses focusing on this area. For larger installations, or for consultancy purposes, employing someone with this level of skill has obvious advantages.

4.3 Maintenance safety

It is important that maintenance works do not put life or property at risk. All maintenance works, whether under emergency conditions or not, must be properly planned and implemented through a safe system of work.

HSR25: Memorandum of guidance on the Electricity at Work Regulations 1989 is available for download from http://www.hse.gov.uk/pubns/books/hsr25.htm. It is strongly advised that the reader obtains a copy of *HSR25* as it highlights the precautions needed, in general terms, to achieve the high standards of electrical safety required by the Electricity at Work Regulations.

Also available from http://www.hse.gov.uk/pubns/books/hsg85.htm is HSG 85 *Electricity at work: Safe working practices*. HSG 85 provides advice on working on a live or dead (isolated) electrical system. It makes the statement that:

> *Most electrical accidents occur because people are working on or near equipment that is:*
> - *thought to be dead but which is live;*
> - *known to be live but those involved do not have adequate training or appropriate equipment to prevent injury, or they have not taken adequate precautions.*

When working on electrical installations it is advisable that an assessment is made as to whether it is absolutely necessary to work live:

(a) does the installation need to be live when testing is carried out?

(b) can it be isolated to provide a safe controlled environment for the maintenance team to work?

Other guidance in this area, such as HTM 06 02 *Electrical Safety Guidance for low voltage systems*, places restrictions on what type of work may be carried out live and what precautions are necessary (see section 8.1 of HTM 06 02). As well as national and industry guidance in this area, which can be generic in nature, local electrical safety policy documents should also be consulted. These should have been compiled to take into account the specific requirements of the local installation, including interlocks, dual feeds, standby supplies and battery back-ups.

There may well be circumstances where operatives have to investigate failures by fault finding whilst equipment is 'live'. Before such activity commences it is essential to ensure that such operatives:

(a) have the correct level of experience;

(b) have undertaken the appropriate training; and

(c) can therefore be classed as 'competent' to undertake live working.

If the work has to be done whilst the main isolator is switched on, then:

(a) a full risk assessment must be done;

(b) the operatives must be fully briefed;

(c) appropriate PPE must be worn; and

(d) correct tooling must be provided.

Whilst undertaking this work, tools, test leads and test equipment must be of the correct 'class' to be safe to use in these circumstances.

It is also advisable for any operative working live on an installation does not do so alone. They should be accompanied by someone who has received first aid training, especially with respect to electrical shocks.

The following diagram is published within HSG 85. The regulations referred to are part of the Electricity at Work Regulations; the paragraph references are guidance within HSG 85 itself.

▼ **Figure 4.1** HSG 85 – Working dead or live

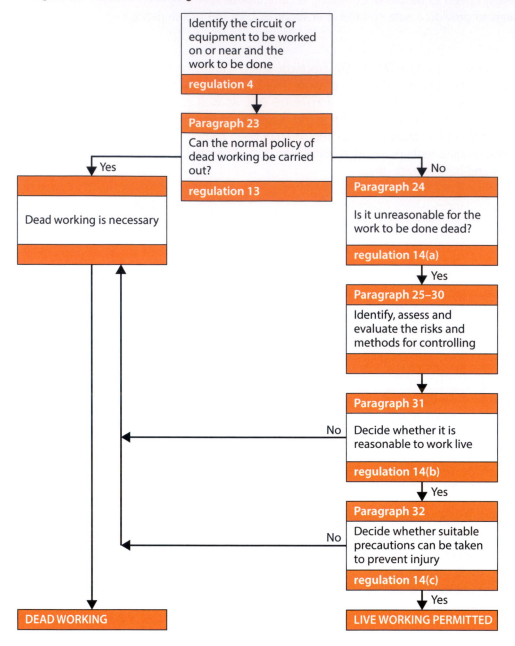

Publications such as the IET *Code of Practice for Electrical Safety Management* provide more comprehensive coverage of safe systems of work.

It should be expected that the following areas should be covered as a minimum as part of the maintenance cycle:

(a) RAMS;
(b) personnel briefing;
(c) prove safe and isolated environment with safety locks applied as necessary;
(d) permits to work and associated documents completed and issued;
(e) safe reinstatement and sign offs; and
(f) return to service.

Safety signage is an important requisite for all electrical installations. More information on permanent and temporary safety signage can be found within BS 5499-10:2014 *Guidance for the selection and use of safety signs and fire safety notices.*

The Health and Safety (Safety Signs and Signals) Regulations 1996 implement the European Council Directive 92/58/EEC on the minimum requirements for the provision of safety signs at work. A download is available from the HSE website (*Safety signs and signals. The Health and Safety (Safety signs and signals) Regulations 1996. Guidance on Regulations*).

Use of temporary signs to denote diversionary routes or additional labels to reinforce information on temporary shutdowns of electrical equipment are an important part of maintenance safety. Whilst safe systems of work and management of risks is the first step, use of additional temporary safety signage complying with the relevant regulatory guidance can assist with mitigating residual risk.

4.3.1 Health and Safety RAMS

Each and every electrical maintenance activity needs to have a health and safety risk assessment carried out on it. In conjunction with this there should be a method statement.

The risk assessment needs to carefully consider each step of the process to ensure that the risks have been properly recognised in order of likelihood and severity and that the appropriate mitigation plans have been put in place.

The mitigation must either remove the risk completely or reduce the risk to an acceptable level that does not place operators, staff or the general public in danger. Similar considerations in terms of risk to property also need to be carried out. The mitigation must be subject to constant monitoring throughout the particular maintenance process to ensure the validity of the risk assessment.

The five principal steps of a risk assessment are:

(a) look for hazards in the electrical maintenance process or procedure (for instance, what will be the impact if services are isolated);
(b) assess who or what may be harmed and how (this may include other building services or equipment);
(c) evaluate the risks and assess if the existing precautions are sufficient or need enhancing (for instance, standby or temporary supplies for critical equipment);
(d) record the findings to demonstrate understanding of the task and for audit purposes; and
(e) periodically review the assessment and update as required - this should be done by operators and supervisors throughout the task.

A method statement will provide a narrative of the work to be carried out, and should give specific details, such as:

(a) the current status of the equipment;
(b) a description of the work to be carried out, including safety measures;
(c) the expected outcome of the work and the new status of the equipment;
(d) timings for the activity;
(e) personnel required;
(f) resources to be used;
(g) PPE required;
(h) permits and other authorisations required;
(i) access arrangements required, including security;
(j) warning signs and barriers that need to be in place;
(k) any necessary diversions for other staff and the general public;
(l) procedures for monitoring and enforcing the warning signs, barriers and diversions;
(m) available first aid facilities;

(n) details of the nearest hospital emergency facility;

(o) the sequence of any electrical isolations;

(p) impacts on services and any standby arrangements;

(q) areas that are off limits and outside of the scope of works; and

(r) the reinstatement plan.

Live working on electrical systems should be avoided where possible. If it is necessary to work on, or near, live parts then all necessary precautions should be taken. Only fully trained personnel wearing appropriate PPE and equipped with the appropriate tools and instruments should be allowed near live parts.

4.3.2 Operational risks

In addition to health and safety risks there will also be operational risks. Electrical maintenance invariably involves the isolation of a system at some point. For operational reasons that may not be a simple task because of the load that is connected, which needs continuous electrical supplies for reasons of life safety, or because of high value commercial risk.

In the following example, within table 4.1, the loads and the risks of a switchgear replacement project are analysed. Although this example applies to a hospital, it could equally be applied to a data centre with various cabinets from multiple clients, each requiring varying degrees of infrastructure support and commercial drivers. Alternatively, it could apply to a large factory complex with critical process lines that could put production at risk if the electrical supply is interrupted, however briefly that may be.

Hospital Substation Operational Risk Analysis

Risk Categories:

HIGH   = Loads with risk to life

MEDIUM   = Loads that disrupt the normal function of a hospital building

LOW   = No discernible risk to life and limb or normal functioning of buildings

▼ **Table 4.1** Example of electrical load operational risk analysis

Circuit designation	Loads	Operational risk category	Risk mitigation measures for circuit changeovers
Research Building	Various – affecting medical experiments	MEDIUM	Careful coordination of shutdown times with end users to avoid disruption of important medical experiments. Use of main and back-up temporary generation and auto-changeover panel to reduce risk of extended supply interruption in the event of generator failure.
LV 5.0 – MCC	• Pumps for medical gases • Air handling plant • Computer room extract fans supplies	HIGH	Out of hours pre-arranged shutdown for changeover.
LV 6.0 – Operating Theatre Switchboard	Lighting and small power to ALL theatres	HIGH	Coordinate with other operating theatre shutdowns. Temporary supplies to be provided from 'A' side.
LV 7.0 – Bore Hole Pump & Water Treatment Plant	Bore hole extraction pump and water treatment plant for bore-hole water	LOW	Changeover during evening shutdown when water demand low. Water mains supply changeover occurs when tank low level switch operates. This initiates solenoid operated valve to open mains water supply.
LV 8.0 – Imaging Machine no1	Essential supplies to angiography	HIGH	Limit supply interruption to short switching transition at start/end of shift changeover.
LV 9.0 – Imaging Machine no2	Essential supplies to angiography	HIGH	Limit supply interruption to short switching transition at start/end of shift changeover.
LV 10.0 – MCC	• Basement plant room • Medical air compressors • HWS circulators • Condensate pumps	MEDIUM	Alternative temporary supply to be provided. Maximum interruption time for changeover is 15 minutes before risk of air receiver pressure decaying.
LV 14.0 – East	• Riser A essential • Riser B essential (includes one SA compressor) • Fire hose reel pumps • Op Theatre Plant	HIGH	Limit supply interruption to short switching transition at start/end of changeover. Ensure other surgical air (SA) compressors are in service prior to start of shutdown.
LV 15.0 – West	• Riser C essential • Riser D essential • Op theatre plant MCC • West lifts A & B	HIGH	Limit supply interruption to short switching transition at start/end of changeover.
LV 16.0 – CHILLERS	2 banks of chillers	MEDIUM	Sequential changeovers with one set always available.
LV 20.0 – PACS CHILLER 1	Imaging chillers	LOW/ MEDIUM	Share load with PACS chiller 2 during shutdown.
LV 21.0 – PLANT	• Private Wing mech services • Includes one SA compressor • Chilled water pumps	MEDIUM	Shutdown, out of hours. Ensure other 2 SA compressors are in service prior to start of shutdown.
LV 22.0 – LS LIFT	Private Wing Lift	LOW	Shutdown, evenings or weekends – alternative lifts available.
LV 25.0 – East NE	• Riser E non-essential • Riser F non-essential	MEDIUM	Limit supply interruption to short switching transition at start/end of changeover.
LV 26.0 – West NE	• Riser B non-essential • Riser D non-essential • Hydraulic goods lift	MEDIUM	Limit supply interruption to short switching transition at start/end of changeover.

4.3.3 Personnel briefing

It is important that once the risk assessments and method statements are completed, the maintenance personnel are properly briefed by their maintenance supervisors or managers on all safety matters pertinent to the task in hand. It is not sufficient for an office-based engineer to simply issue paperwork to the maintenance personnel and let them get on with it.

The maintenance personnel should demonstrate that they understand the requirements of the task and should acknowledge that by signing the method statement to accept their own responsibilities under the terms of the Electricity at Work Regulations.

It should be recognised by all that this action of signing by the operators is not a means of passing ultimate managerial responsibility to them but a means for them to appreciate the risks and the agreed mitigations.

4.3.4 Prove power sources are safe and isolated

It is safer and preferable to carry out any electrical maintenance with the power source switched off and isolated. The work itself will typically be carried out by members of the maintenance team who will be recognised locally as a competent person (CP). However, prior to the work commencing, the isolations must be carried out either by the local authorised person (AP) or under their direct supervision. The purpose of these isolations is to prove the electrical system is safe and isolated with safety locks applied as necessary.

The AP must then demonstrate that the system is safe to work on before handing over the equipment to the CP. The CP should not accept the electrical system for maintenance unless this process of isolation and demonstration of safety takes place.

Once this is complete safety lock systems with twin locks should be used, with one key held by the CP and one by the AP. In the event that there are several contractors involved each contractor must be provided with the facility to be assured of a safe isolation. This can be achieved by the use of multiway lock-off hasps together with isolation warning labels.

4.3.5 Permits to work and associated documents

Occasionally it will be necessary to make safe large parts of the installation to undertake some maintenance procedures on a relatively small part. In such an instance, in addition to the preparation of a method statement, switching procedures and isolation and earthing diagrams, permits to work are also necessary administration that must be completed by the AP.

Copies are then issued to the CP. The CP shall retain the copy until the work is complete and handed back. The copies of the permits to work and associated documents should then also be handed back to complete the paperwork trail. This activity then needs to be recorded in the site logbook.

4.3.6 Safe reinstatement and sign offs

It is important that, once the electrical maintenance work is satisfactorily completed, close control of the maintenance process is also carried through to the reinstatement of the service.

As part of the method statement process above any required reset procedures should also be thought through and noted. This is especially important if electrical isolations have been required, particularly in critical locations such as hospitals or commercially sensitive areas such as data centres. Interruptions to electrical supplies and services need to be kept to a minimum, but reinstatement needs to be made in an orderly fashion to ensure that it works first time. Repeated interruptions should be avoided.

The purpose of a reinstatement procedure is to set all services that may have been affected by the electrical maintenance back into service. In this instance, it will be advantageous to prepare a comprehensive schedule of all services that need checking as an aide memoire. The preferred operational status of affected services should be noted as some services will need to be in standby mode rather than run mode.

Where necessary, checklists and schedules should be signed off by the appropriate personnel to demonstrate that the maintenance team has satisfactorily handed back control. Any permits will need to be cancelled as part of this process.

4.3.7 Return to service

Following completion of both the original electrical maintenance work and the associated administrative paperwork, the maintenance personnel should then tour the areas affected by any electrical isolations to ensure that all end users are satisfied that the systems are working satisfactorily and that all user interfaces are operating satisfactorily.

Once this is done the maintenance cycle is almost complete.

4.3.8 Waste disposal

An inevitable by-product of some electrical maintenance activities is waste, which may range in size from small redundant components to larger complete electrical equipment. The components will typically be time expired and replaced as part of the maintenance tasks. The electrical equipment may be beyond economical repair.

It is beholden on the maintenance team under various items of legislation to dispose of this specialist waste responsibly and through the appropriate channels.

The reader may wish to review the following:

(a) Environmental Protection Act 1990 Section 34 *(Waste Management – the Duty of Care – A Code of Practice)*
(b) The Waste Electrical and Electronic Equipment (WEEE) Regulations 2013
(c) Scope of equipment covered by the UK WEEE Regulations (LIT7876)

Most local councils have schemes to enable the safe disposal of electrical and electronic waste for householders. Information on business waste disposal should be carefully researched to ensure correct disposal. The WEEE Regulations also calls for manufacturers to provide proper disposal routes for such waste. Government-produced guidance notes can be found at: https://www.gov.uk/government/publications/weee-regulations-2013-government-guidance-notes

When disposing of waste, a duty of care for safety as well as the environment is paramount. For example, old lamps, such as linear and compact fluorescents, are prone to exploding, leaving dangerous glass shards. Ideally specialist companies should be used to remove waste lamps from site.

Evaluation of your electrical system

5.1 General overall pattern of maintenance evaluation

In Section 2.4 the typical pattern of electrical maintenance was discussed. There are statutory requirements on some electrical systems, such as emergency lighting and fire alarms, for periodic testing and functional checks. However, as with any properly managed process, there ought to be a feedback loop on all electrical maintenance regimes to improve the process.

Electrical maintenance is a process, depending on the system, that is repeated several times through an installation's lifecycle. A properly managed maintenance regime is tailored and adjusts to suit the age of the installation. It allows judgements to be made on replacements of systems and gives guidance on how much longer a system can last to assist with capital project planning. When fairly new and just out of warranty, an installation in a relatively benign environment will require less attention and component replacements. A few years down the line the intensity of the maintenance regime must increase or the installation should be replaced.

One fairly simple, and widely recognised, cycle of activities, which fits neatly into any maintenance regime, is 'Plan, Do, Check, Act'.

Within Section 2.4 the first three elements were covered in more detail. Appendices B1 to B4 inclusive will provide practical examples of particular parts of an electrical installation covering all three of these elements. Figure 5.1 demonstrates how these first three elements integrate into the Plan, Do and Check parts. The last stage, Act, which is largely management and administrative based but is no less important, is explored further below:

▼ Figure 5.1 Plan, Do, Check, Act – the maintenance cycle

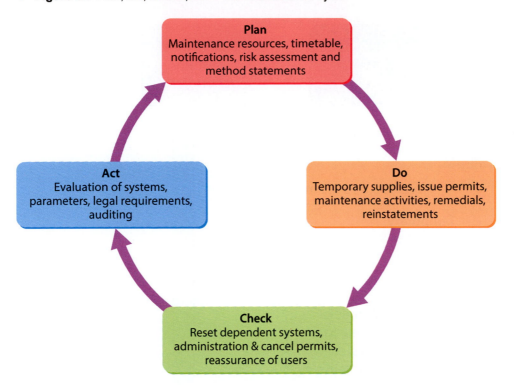

Exploring this idea and developing the original pattern of activity allows for the addition of two further activities:

(a) careful evaluation of the electrical installation; and
(b) an auditing trail of a maintenance regime.

The first point (a) eases the process of preparation without, for instance, the time constraints or pressures caused by reactive maintenance. The second point (b) is used to demonstrate compliance with statutory requirements.

This part of the process should not be underestimated: it provides close control of an installation and plays a significant role in extending the life of an electrical installation. It also provides documented evidence if an event needs to be investigated by HSE or another body.

Using figure 3.1 Section 3.4 the evaluation and auditing components can be integrated into the wider process, hence showing how the maintenance process feedback loop can be closed:

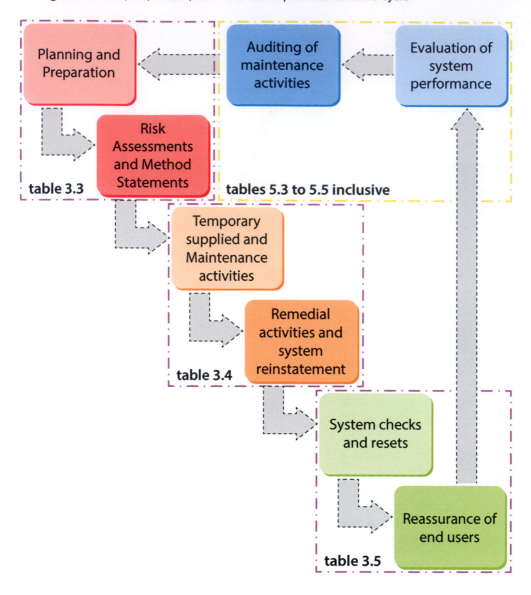

5.2 Evaluation of electrical systems

Using the O&M manual as a starting point, the electrical installation should be completely evaluated to record exactly what systems and equipment are installed.

Beyond the obvious, such as lighting and small power, other support services such as emergency lighting; fire detection and alarms; security such as closed circuit television (CCTV); and access control all need to be noted. Heavier infrastructure items, such as electrical switchrooms, lift installations and mechanical plantrooms, need to be included too. Newer technologies, such as solar panels, DC distribution systems and energy storage should not be ignored.

A picture can then be built up by the electrical maintenance engineer for each system using some basic and more detailed questions. For instance:

(a) the nominal date of commissioning (i.e. age of the installation);

(b) the performance criteria at commissioning;

(c) the nominal lifecycle of the equipment or system – manufacturer's data;

(d) the manufacturer's recommendations on maintenance frequency;

(e) statutory requirements on the types of functional checks;

(f) statutory requirements on periodic testing;

(g) the operating environment – how harsh is the temperature, humidity, water ingress, dust etc.;

(h) the availability of replacement components – leading to questions on obsolescence;

(i) importance to safety of occupants;

(j) importance to business continuity;

(k) importance to building integrity; and

(l) importance to energy performance of the installation.

Answers to these queries will be fundamental to informing the maintenance strategy for each electrical system. *It is important to remember that one size does not fit all.*

Varying maintenance strategies will be required for different electrical systems and also for different parts within an electrical system. The local operating environment should also be considered, for example, the lighting in a boiler plantroom will need a different approach to lighting in a main access corridor near offices or within a meeting room.

Some systems will need to have a functional test carried out regularly to ensure that the system will operate satisfactorily when required. Various industry standards and local policy documents will determine the frequency of this, be it weekly, monthly, quarterly or annual.

In addition, there will be a need for periodic verification tests to ensure that the condition of the electrical system still meets the required installation standards. Invariably this will require certification to be issued, and this is often coupled with remedial activities. These may be completed annually or perhaps every five years.

Some systems may only be verified every 10 years. Such decisions are based on tables within a variety of industry standards and should be tempered with knowledge of the local environment and frequency of use.

Table 5.1 explains some of the typical types of simple functional tests and status verification tests that may be carried out on an electrical installation. Table 5.2 develops this further to show how those tests will relate to different types of electrical installation. This list is an example only and should not be considered definitive. The reader is encouraged to verify the frequency of tests with the relevant standards and also to assess that against the environment in which their own installation is situated.

▼ **Table 5.1** Types of test and frequency

Code	Type	Explanation
FT	Functional test	• simple on/off test to ensure the system operates satisfactorily.
VT	Verification test	• comprehensive check to ensure the viability of the system and the performance characteristics.
W	Every week	• this is an activity that is expected to be carried out weekly. • results usually logged by tick box approach.
M	Every month	• this should be a monthly exercise. • results to be logged by tick box approach, perhaps with additional comments where necessary.
Q	Every three months	• this test should be undertaken four times a year (i.e. quarterly). • results to be logged by tick box approach, usually with additional comments.
A	Annually	• a test and inspection carried out once a year. • results to be logged on a check sheet, usually with additional comments where remedials are required. • certification will also be issued for compliant systems.
5Y	Five yearly cycle	• a full test and inspection carried out once every 5 years. • results to be logged on a check sheet usually with additional comments where remedials are required. • certification will also be issued for compliant systems.

An electrical installation may include some or all of the following systems. This list of examples is not exhaustive and the reader is encouraged to examine their own installation to include other items that are not on this list.

The frequency and type of testing may also vary according to the manufacturer's own recommendations. The period between testing should be reduced if the installation environment is deemed to be harsh or altered from the original design intent.

Table 5.2 Electrical installation schedule – test and frequency

Electrical Installation Schedule	FT	FT	FT	FT	VT	VT
	W	M	Q	A	A	5Y
a.c. distribution, switchboards and switchgear	♦			♦		♦
a.c. sub-distribution boards and sub circuits	♦			♦		♦
Lighting			♦			♦
Emergency lighting	♦				♦	
External lighting				♦		♦
Fire alarms	♦				♦	
Gas detection systems	♦				♦	
CCTV				♦		♦
Access control	♦					♦
Disabled person alarms (toilets)	♦					♦
Nursecall systems (Healthcare)	♦					♦
Mechanical plantrooms			♦			♦
Controls			♦			♦
Leak detection systems				♦		♦
Pipework trace heating				♦		♦
Electrical switchrooms			♦			♦
Earthing				♦		♦
UPS			♦			♦
Isolation transformers				♦		♦
Electrical heating		♦			♦	
Standby generation		♦				♦
Photo-voltaic panels				♦		♦
Wind turbines			♦	♦	♦	
d.c. rectification, sub-distribution and sub-circuits				♦		♦
Data hub rooms				♦		♦
Lifts				♦	♦	
BMS/BEMS		♦			♦	

5.2.1 Assessment of system status

As shown in table 5.3, a number of items need to be checked and assessed to establish the status of a system and provide the initial step in ensuring the correct maintenance programme is adopted.

Survey	Has the system been physically reviewed with a non-intrusive survey to note the status of any deterioration since commissioning?
	(a) components failures e.g. lamps out etc.
	(b) component damage e.g. impacts on corridor sockets from trolleys.
	(c) surface deterioration from humidity, extremes of hot and cold.
	(d) changes to the building layout which inhibits access arrangements.
Commissioning data	Do the O&M manuals record:
	(a) as-built commissioned settings for relevant equipment?
	(b) as-built layout drawings?
	(c) as-built schematics?
	(d) details of recent upgrades/alterations/refurbishments?
	(e) impacts for updates and redundant systems?
Operational performance	Is there a document or operational criteria that makes clear:
	(a) the acceptable minimum and maximum operational level for monitoring purposes?
	(b) the baseline of operational performance?
	(c) a record of usage for run and standby systems? Is the record similar for both?
	(d) a mimic diagram to demonstrate the current status of switches and lock offs for complex distribution systems with system bypass, bus-couplers and interlocks?
Intended use	The intended use and expected outcomes of the electrical system should be evaluated against the:
	(a) age of the equipment;
	(b) existing condition of the equipment;
	(c) duty cycle;
	(d) harshness of the environment in which it operates;
	(e) capability of users (is it subject to operator abuse?); and
	(f) experience and training of the user.
Maintenance resources and timings	Determine the approach to maintenance for a particular system:
	(a) fully trained personnel or supervised non-technical personnel?
	(b) in-house staff or outsourced contractor/specialist?
	(c) lead time for replacement components/systems?
	(d) timing of activities e.g. out of hours/holiday periods etc.?

5.2.2 The manufacturer's information

The manufacturer's data is critical to enabling the efficient maintenance of an electrical system.

Relevant equipment data should have been provided in the O&M manual. It is important that this does not contain endless reams of generic catalogue references for the equipment that has been installed. As table 5.4 suggests, this information should be specific to the actual site it is installed on where necessary, with any bespoke adjustments and amendments noted.

Manufacturer's instructions	Determine manufacturer's maintenance instructions for the relevant equipment. This may include: (a) diagrams; (b) diagnostic charts; (c) schedules of components; (d) models; (e) performance criteria; (f) programming lists; (g) engineers codes and reset data; and (h) other reference material.
Layout drawings	• Check site specific layout drawings as issued by the manufacturer's installers. • These should be 'as installed', but may in reality just be copies of the working drawings used during installation, or even the original design drawings. • Are they accurate and complete?
Schematic drawings	• Check site-specific drawings as issued by the manufacturer's installers. Again these should be 'as installed'. • It is vitally important that any on-site changes or adaptations made during commissioning are correctly noted to enable diagnostic checks and intrusive remedial works to be carried out correctly.

5.2.3 Evaluation of equipment failure modes

If the system works and then it fails:

(a) have the consequences of that failure been thought through already?
(b) have the risks to life and property been discussed?
(c) is a commercially sensitive process put in jeopardy by the failure?
(d) how quickly can the system be restored?
(e) are there mitigations in place, such as alternative systems or alternative electrical supplies?

Table 5.5 explores these themes.

Impact on safety of system failure	Evaluate and assess the risk to safety: (a) what are the safety consequences of failure – are people or property at risk? (b) are there contingency plans in place? (c) what is the criticality of failure in terms of downtime?
Impact on business of system failure	Evaluate and assess the risk to the business: (a) what are the business continuity consequences of failure? (b) are there contingency plans in place? (c) what is the criticality of failure in terms of downtime?
Potential failure modes	Evaluate the relevant failure modes of the equipment. This could be based on: (a) previous experience; (b) the manufacturer's information; (c) trade association information; and (d) peer networks etc.
Collation of data	If regular failures are occurring on a particular electrical system it is important to collect data from those failures: (a) are there patterns that dictate why these failures happen? (b) can those patterns be altered to improve the longevity of the systems between failures? (c) are there particular issues relating to location of a luminaire, an insect infestation causing a smoke detector to activate, or prevailing weather affecting a street lamp?
Changing behaviour	Following an evaluation of the results of regular failures it should be possible to change the pattern to ensure fewer failures. This may require: (a) a change in behavior of the people using a particular part of the installation; or (b) using more robust replacement components; or (c) increasing the observational part of a maintenance programme to be more prepared for failure when it occurs.

5.2.4 Implementation and planning

With the information gathered and evaluated the next step is to plan and implement. Different approaches will be required depending on the maintenance scenario. However, it should be possible to put into place beforehand procedures for all types of maintenance activities even when reacting to unexpected component failures or unplanned system breakdowns. In this way the remedial activity is hopefully more straightforward, reinstatement quicker and any risks are a known quantity.

It is also helpful for staff, especially those less familiar with a particular site, if these procedures are available. Electrical maintenance staff will be skilled personnel, but each site has its own peculiarities as it evolves depending on layouts and applications of the electrical systems.

One way of developing these procedures is to carry out a desktop study of a particular electrical system and envisage a likely scenario that would potentially need to be dealt with.

A collaborative exercise or workshop with principal stakeholders can explore the details of how the electrical maintenance team might deal with the scenario. From this a flow diagram (such as in figure 5.3) can be developed to expand on the likely scenario, questions can be raised and solutions developed by way of preparation (refer to table 5.6 as an example).

Such a scenario may be the failure of a system at 8.00pm on a Saturday night. What should the first line of response be?

▼ **Figure 5.3** Failure scenario flow diagram

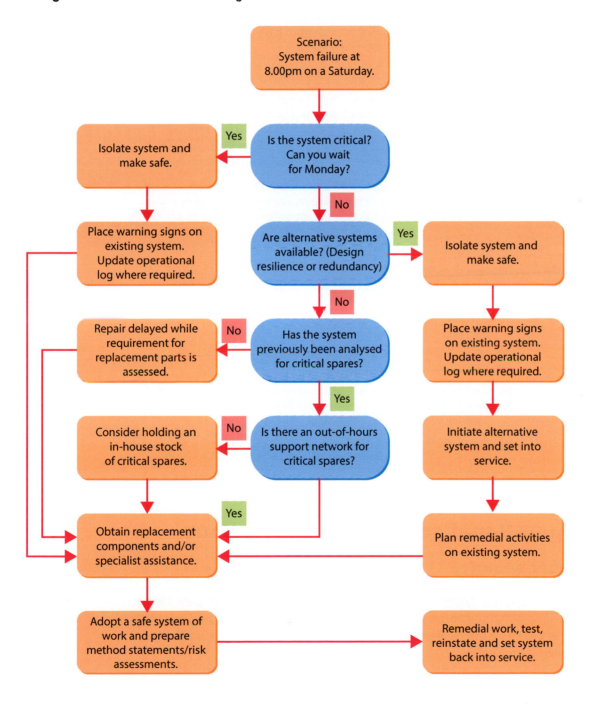

Maintenance strategy for particular system	Determine the approach and activities for the maintenance of the electrical system, the equipment and any replaceable components. (a) does the approach vary according to the type of maintenance required? In a remedial situation there will be additional time pressure to reinstate the electrical system and get it back up and running. (b) are additional checks actually required to ensure short cuts are not made and mistakes do not occur?
Standard operational procedures	(a) As an aid to some electrical maintenance activities it can be useful for more complicated systems to have a pre-written standard operating procedure (SOP). (b) SOPs can act as an aide memoire to ensure accuracy and completeness in the maintenance task and assist safety. (c) SOPs are especially useful when carrying out switching activities on large switchboards, when switching between dual feeds and bus-couplers or on bypass operations for UPS equipment. (d) Usually switching activities involve the use of interlock keys such as Castell or similar, where there may be an element of confusion.
Holding of spares and other resources	Clearly it may be advantageous for some spare components to be held in-house to assist with first line maintenance and improve response times; however, this policy is not without questions of its own: (a) what are the capital costs of such items? (b) how does that compare with the costs and benefits of holding them against the delayed reinstatement of the system whilst spares are sourced? (c) is there a shelf life associated with the spare components? Will they perish before they get used? For instance, the shelf life of fuel within bulk fuel tanks for standby generators should also be assessed.
Programme planning	(a) Setting aside the issue of reactive maintenance, generally, with all information evaluated, it will be possible to implement an agreed electrical maintenance programme. (b) It is important to record all activities and the outcomes so that any remedial tasks can be managed effectively and subsequently signed off as complete.

5.3 Operational security considerations

5.3.1 Cyber security

Increasing use of computer-based systems for the management of maintenance tasks and the use of web-based software brings into play the spectre of cyber security, and the associated risks that this issue raises. It is, however, beyond the remit of this publication to deal with this in any detail.

In this area the IET acknowledges two particular challenges:

(a) the need to improve cyber security awareness and the skills of those working in the engineering and technology sectors; and
(b) the development of guidance material about the design of cyber security for intelligent buildings.

More information can be found on the IET website. Building on the previous release of a briefing in this area, the *Code of Practice for Cyber Security in the Built Environment* has been published by the IET; see http://www.theiet.org/resources/standards/cyber-cop.cfm.

Additionally, BSI, through their publically available specifications (PAS), promotes the development of good working practices in cyber security as the use of BIM becomes more widespread.

The need for careful attention to cyber security is apparent throughout the life span of an installation as more and more information is placed on servers accessed through the wider internet, either locally installed behind company firewalls or remotely on a cloud server. This attention has to be followed diligently at all stages from concept design, through design development, construction, commissioning and handover. The need for continuing security throughout the operational period and the maintenance of an installation is also apparent.

Increasingly, maintenance monitoring will become part of the Internet of Things as all manner of devices, systems and controls are directly connected to data networks, and as this develops the likelihood of a weak link increases – this could then be exploited. Cyber security, historically in the realms of science fiction, is already a fact.

As maintenance managers come to rely more heavily on cloud-based programmes to manage maintenance programmes via a data network connection, the need to ensure that an installation is secure from cyber exploitation becomes a much greater priority.

The IT infrastructure that supports such maintenance activities needs to be carefully considered, including dedicated servers for as-built documents and ongoing maintenance records. Where these servers can be accessed via the internet they may need protecting behind dedicated firewalls and associated IT protective strategies. Organisations with extensive use of information technology systems may benefit from the adoption of, and where appropriate, certification to, BS ISO/IEC 27001:2013 *Information technology. Security techniques. Information security management systems. Requirements.*

Perhaps an unrecognised issue for exploitation with networked equipment is the facility for 'factory defaults' or 'engineers' codes' on networked equipment. Another issue perhaps more widely acknowledged is that of departing staff being a weak link in security.

Where equipment is connected to data networks, an overall security strategy should be established to ensure that data and equipment cannot be accessed from the internet by known means. In particular, default settings, machine names, user accounts and passwords provided with equipment as default should be considered wholly unacceptable.

Regular changes of password should be implemented, and the strategy should include for individually allocated user accounts as far as practicable, to ensure user access is disabled at appropriate times (e.g. staff turnover).

5.3.2 Document security

The compilation of SOPs, complete with diagrams and photographs of the equipment, is a useful tool in maintenance especially when it comes to complicated switching arrangements. Personnel carrying out the switching are experienced and usually assessed directly on the switchboards and similar equipment they are faced with.

However, it is often beneficial to offer an aide memoire to minimise mistakes. Interlocks should prevent dangerous occurrences but switching from one source of supply to another may cause confusion, whether it is done via bypass arrangements or reinstatements.

Having provided the SOP it is often prudent to consider where to locate these documents. Clearly a copy should be within the main O&M manual; however, some maintenance teams may deem it useful to leave a copy within a switchroom, for example, for ease of reference and perhaps displayed next to a schematic of the wider electrical network.

Depending on the sensitivity of the installation other teams might deem that leaving such information to hand for unauthorised persons to view is a security risk. In this case any SOPs would be held centrally where they can be issued at time of need only.

5.4 Monitoring of remedial activities

It is important not to lose sight of remedial works that were deemed as a lower priority during the principal maintenance works. These should be managed in a suitable timeframe to ensure that they are completed satisfactorily.

Some larger organisations use a maintenance backlog system to manage these items. Where items cannot be completed within an agreed period they are usually placed on the site risk register. From there the 'at risk' system can be monitored at managerial level with input from capital planning engineers and project managers, compliance officers and also maintenance teams. This strategy would work for systems that have been earmarked for replacement in a future financial period where spending more maintenance budget could be considered not viable.

5.5 Auditing of maintenance procedures

It is advisable that all maintenance activities are audited from time to time to ensure compliance with statutory requirements. BS EN ISO 19011:2011 *Guidelines for auditing management systems* is the accepted standard for auditing management systems, includes requirements for defining and assessing the competence of auditors, and is recommended for all audits. Further industry-specific guidance is available, as discussed in this section.

For example, within the healthcare sector, in documents HTM 06 02 and HTM 06 03 the roles of both CPs and for APs are described so that day-to-day operational maintenance and management requirements can be fulfilled. Their roles are clearly defined, with the AP overseeing the CP and reporting in to a more senior manager or duty holder.

However, the documents also describe a role for an authorising engineer (AE) to oversee the safe systems of work and periodically audit maintenance activities to ensure that compliance with Health and Safety legislation has been achieved. Typically, an annual review of the maintenance procedures takes place with records checked. Site surveys can reveal a good deal of information on the diligence of the maintenance teams and on how robust their procedures actually are. The competence levels and qualifications of the maintenance team can also be monitored in this way.

The MOD has a similar role with its JSP 375 publication and the role also exists within the data centre industry at larger organisations with an extensive portfolio of properties.

Ideally an auditor should check that the maintenance carried out complies with the statutory requirement of the Electricity at Work Regulations and supporting documents. The auditor needs to check that life-safety systems have been regularly tested and that any remedial actions recorded have been acted upon.

In addition, an annual report should be compiled to:

(a) evaluate the maintenance regime against recognised criteria;
(b) confirm that the installation complies with legislation in terms of safety; and
(c) provide the duty holder with recommendations for improvement where required.

Spot checks on the actual electrical installation should also be carried out to ensure switchrooms and electrical risers are kept in a good state of repair, clear of rubbish and with appropriate safety measures, signage and security.

Within HTM 06 there is a requirement that such an auditor should be independent of the local management to ensure that any critique is fair and balanced, and not swayed by local influences. The AE reports directly to the management.

APPENDIX A

Checklists

The following checklists are based on items discussed within sections 1, 2 and 3. They are set out here to provide example proformas to use when collating information to assist with maintenance activities.

These lists are not presented as definitive and an analysis on site may reveal more items to add to the lists; some items may even be removed. There are tick box columns to record the applicable status to the questions posed and simply represent Yes (Y), No (N) and Not Applicable (N/A). There is a comment column to record details and note outcomes.

Appendix A1 focuses on the impact that design can have on maintenance. When using these checklists it is important to track changes and to note where the maintenance stakeholders can influence design. It will also be important to note where the design has not been altered and to plan ahead for the impact that that result may have on maintenance tasks going forward.

▼ **Appendix A1** Design for maintenance checklists

Refers to section:	Topic for assessment	Checklist table reference
2.1.2	Maintenance stakeholders	Table A1.1
2.1.4	Space planning	Table A1.2
2.1.5	Maintenance resources	Table A1.3
2.1.6	Component selection	Table A1.4

Appendix A2 focuses on the general activities that are required immediately before, during and immediately after any electrical maintenance tasks. Paying heed to the criteria in these tables, coupled with the safety concepts demonstrated in the IET *Code of Practice for Electrical Safety Management*, will provide the framework for a properly managed maintenance regime.

▼ **Appendix A2** Maintenance activity checklists

Refers to section:	Topic for assessment	Checklist table reference
3.4.1	Before the event	Table A2.1
3.4.2	During the event	Table A2.2
3.4.3	After the event	Table A2.3

Appendix A1 Design for maintenance checklists

▼ **Table A1.1** Maintenance stakeholders
Refers to the text within section 2.1.2

Criteria	Y	N	N/A	Comments
Operational restrictions				
Equipment preferences				
Market availability				
Specialist training				
Local standards				
Derogations				
Downtime				

▼ **Table A1.2** Space planning
Refers to the text within section 2.1.4

Criteria	Y	N	N/A	Comments
Design and space planning				
Future expansion				
Manufacturer's guides				
Safe access and egress				
Structural considerations				
Manual handling				

▼ **Table A1.3** Maintenance resources
Refers to the text within section 2.1.5

Criteria	Y	N	N/A	Comments
Training at technician level				
Training at managerial level				
Safe systems of work				
Manual or automatic monitoring				
First line response				
Compliance checks				
Specialist equipment				

▼ **Table A1.4** Component selection
Refers to the text within section 2.1.6

Criteria	Y	N	N/A	Comments
Value engineering				
Quality				
Economy				
Backwards compatibility				
Obsolescence				
System upgrades				
Stakeholder consultation				

Appendix A2 Maintenance activity checklists

▼ **Table A2.1** Before the event
Refers to the text within section 3.4.1

Criteria	Y	N	N/A	Comments
Training and familiarisation				
Out-sourcing				
Contingency planning				
Resources				
RAMS				
PTW, safe systems of work and logbooks				
Notifications to end users				
Notification to other occupants				

▼ **Table A2.2** During the event
Refers to the text within section 3.4.2

Criteria	Y	N	N/A	Comments
Task briefing & management				
Standby & bypass supplies				
Temporary supplies				
Electrical Isolations				
Maintenance				
Management				
Remedial (system critical & urgent)				
Reinstatement				

▼ **Table A2.3** After the event
Refers to the text within section 3.4.3

Criteria	Y	N	N/A	Comments
Housekeeping				
Administration, permits etc.				
Operator debrief & wash up				
Reassurance of end users				
Follow-up checks				
Review				
Remedial (non-critical and less urgent)				

Suggested maintenance activities

As discussed previously, the Construction (Design and Management) Regulations 2015 (CDM Regulations) places responsibilities on specific duty holders to ensure the documents handed over with a completed installation, as part of the O&M manual, provide all the necessary information required to conduct maintenance.

CDM 2015 Regulation 9 (4) states that:

> A *designer must take all reasonable steps to provide, with the design, sufficient information about the design, construction or maintenance of the structure, to adequately assist the client, other designers and contractors to comply with their duties under these Regulations.*

Information on particular maintenance activities should be taken from the installation's own O&M manual and from manufacturers' details in the first instance. This should have been compiled in accordance with the duties of the design and construction phase contractor as laid down in the CDM Regulations.

If correctly compiled, at the time of project completion, the O&M manual should contain specific manufacturers' guidance on the maintenance of particular equipment and the designer's risk assessments relating to issues with the particular installation. Minimum design (and hence performance) parameters will typically be governed by the British Standards Institution or appropriate local regulations.

The following pages are based on examples of good practice. They provide references and headline maintenance activities for specific types of electrical installation. They are not intended to be comprehensive or supersede the requirements already laid down within the O&M manuals of the reader's own installation. The readers should use the references to follow up with further research. Many of the tables are based on real world situations and should be adapted to suit the reader's own installation.

Each section contains references to assist the reader.

Appendix B1 – Energy input to electrical systems

B1.1	Main switchgear and switchboards (LV and HV)
B1.2	HV/LV transformers and isolation transformers
B1.3	Generators and standby supplies
B1.4	UPS
B1.5	CHP electrical components
B1.6	Solar photovoltaic installations
B1.7	Earthing
B1.8	Lightning protection
B1.9	Surge protection
B1.10	Power factor correction
B1.11	Harmonic filters

Appendix B2 – General applications and circuit protection

B2.1	Protective devices, overloads, discrimination and grading
B2.2	General lighting
B2.3	Small power
B2.4	Electrical heating of hot water
B2.5	d.c. systems

Appendix B3 – Life safety systems

B3.1	Emergency lighting
B3.2	Fire detection and alarm systems
B3.3	Gas/carbon monoxide detection
B3.4	Personnel alarms and call systems (security, disabled persons, patients)

Appendix B4 – Industrial and control

B4.1	Hazardous areas and equipment
B4.2	Motors and variable speed drives
B4.3	Control panels
B4.4	Machinery
B4.5	Automation, control and instrumentation
B4.6	BMS

Appendix B1 Energy input to electrical systems

Introduction

The activities below provide some specific guidance on various forms of energy input to electrical installations. Complex larger installations can have more than one intake location, usually at HV. However, with the advent of on-site renewable technologies even the smallest of installations could have more than one point of electrical supply. When performing maintenance, care should be taken to ensure that each supply is isolated to ensure back feeds do not cause injuries or worse.

Below, several British Standards are mentioned to provide the reader with further areas for research. Other standards will exist in local regions and many of the more recent BS EN series are actually based on agreed European standards.

Selected highlights of BS 6423 include guidance on isolations before and during maintenance work (4.8), approaches to working on live equipment (4.9), caution with stored electrical energy systems (4.12), battery maintenance (especially for switchboard tripping circuits) (5.12), and monitoring changes to the operating environment (8.2).

BS EN 62305 is a document in four parts covering lightning protection systems. As well as discussing the design and the assessment of risks with regard to lightning protection zones, elements of the document also cover the use of surge protection devices, which is referred to in BS 7671:2008+A3:2015.

Standards

- BS 6423:2014
  Code of Practice for the maintenance of low-voltage switchgear and control gear

- BS 6626:2010
  Code of practice for the maintenance of electrical switchgear and controlgear for voltages above 1 kV and up to and including 36 kV.

- BS EN 50522:2010
  Earthing of power installations exceeding 1 kV a.c.

- BS 7430:2011
  Code of practice for protective earthing of electrical installations

- BS EN 62305-1:2011
  Protection against lightning. General principles

- BS EN 62305-4:2011
  Protection against lightning. Electrical and electronic systems within structures

- BIP 2118:2007
  Protection against lightning. A UK guide to the practical application of BS EN 62305

- BS 7671:2008+A3:2015
  Requirements for Electrical Installations. IET Wiring Regulations

- IET Guidance Note 8: *Earthing and Bonding, 2nd Edition*

B1.1 Main switchgear and switchboards (LV and HV)

HV installations occur on larger sites where it is not economical for the local electricity supplier to provide large LV-rated supplies because of the associated current requirements. By providing an HV supply the current can be reduced for a similar power rating. However, HV supplies and their switchgear bring with them particular safety risks, mitigations, regulations and maintenance procedures. Specially trained resources are also required.

One important item of safety to note is that some HV installations employ the use of fire suppression systems with switchrooms and other enclosed areas. Fire suppression systems can be hazardous to health if they are discharged, especially older designs, as they reduce the amount of oxygen in the air.

If there is a need for maintenance personnel to enter these spaces it is important to disable these systems using the appropriate control mechanism and to reinstate the fire suppression system when exiting.

LV installations at low risk sites will present fewer problems so that most suitably qualified and experienced electrical staff should be able to cope. However, on more complex sites, appropriate training and site-specific control procedures may be required to ensure that supplies to critical loads are maintained as much as possible. This may involve the use of alternative supplies, electro-mechanical interlocks and trapped safety keys.

Clearly, maintenance of these areas needs to be carefully considered. The following two documents will assist with that:

- BS 6423:2014 *Code of Practice for the maintenance of low-voltage switchgear and control gear*

- BS 6626:2010 *Code of practice for the maintenance of electrical switchgear and controlgear for voltages above 1 kV and up to and including 36 kV.*

Some principal activities that can be included are shown in Table B1.1:

▼ **Table B1.1** Switchgear and switchboards (LV and HV) principle activities

	Activity	Frequency
1	Security Check that all switchroom and/or wire enclosures are secure and in a good state of repair.	Monthly
2	Operation Visual check for obvious signs of discolouration, heat spots or smell of burning on electrical equipment/switchgear and faceplates.	Weekly/monthly
3	Operation Check all meters and signal lamps function correctly and replace as required.	Quarterly
4	Operation Check all BMS connections function correctly and repair as required.	Quarterly
5	Operation Visual check that all switches and interlocks are in the correct state – does this correlate with any mimic diagrams/logbooks etc. that may be available? Highlight any discrepancies and await further instruction from the site AP.	Quarterly
6	Operation Visual check that all isolated switches have been satisfactorily locked off and complete with warning labels, especially those feeding areas currently under refurbishment. Check with project teams for likely programme of reinstatement.	Quarterly
7	Housekeeping Check there is an insulated mat in front of all switchgear and that it does not foul the passage of any HV withdrawable switch trolleys.	Quarterly
8	Housekeeping Switchroom not to be used as a store and it should be left in a clean and tidy state. All explosive or flammable materials should be elsewhere in the appropriate enclosures.	Quarterly
9	Thermographic Survey the electrical switchboard with calibrated thermal imaging equipment and check for potential hot spot. A note shall be made that a risk assessment will be required for the temporary removal of safety barriers and/or bypassing of safety measures to gain access to switches. Refer to Guidance Note 3 (section 3.11).	Annually
10	Admin Check that accurate and updated schematics are available and located within all relevant switchrooms and adjacent to switchboards. As well as a periodic check this should also be carried out after any modifications carried out under a capital project scheme.	Annually

B1.2 HV/LV transformers and LV/LV isolation transformers

By definition transformers are machines with no moving parts. On this basis some people put forward the notion that transformers are maintenance free. The reality is that no electrical device is truly maintenance free. Transformers are typically very efficient at converting electrical energy with losses of less than 5 %. These losses dissipate through the cooling mechanisms of the transformer and could have an effect on the heat gain in the immediate area, especially if the transformer is located within an enclosure with little ventilation.

The characteristics of the electrical load could have an adverse effect on the transformer. If the load has high levels of harmonics or resonant characteristics then the transformer windings could be affected. There are also environmental considerations to ensure that humidity and other factors do not corrode the transformer casing or cause insulating oil to break down. Transformers, especially oil-based units, often have monitoring facilities and should have a regular visual inspection by electrical maintenance teams. They should also receive specialist periodic testing and assessment by appropriately trained personnel.

As a general rule HV/LV transformers are usually found on larger installations where the LV supply load requirements are in excess of 400 A per phase (around 280 kVA). In such cases the local electricity suppliers prefer to provide power using HV. Like the HV switchgear, HV/LV transformers also have safety risks, mitigations and maintenance procedures, together with specially trained resources.

LV/LV transformers are used in special locations where there is a need to control the earth fault path. At its simplest this will be a step-down transformer to supply a safety extra low voltage (SELV) to a shaving socket in a bathroom. Maintenance regimes for these applications may only amount to periodic visual inspection for corrosion from the effects of humid bathroom air.

For more complex installations it will be 1:1 LV/LV isolating transformers in hospital operating theatres or intensive care recovery areas. Here there are complex monitoring circuits and control philosophies. Immediate isolation of a circuit on detection of one earth fault may not be desirable because a patient's life may be depending on the apparatus that is connected. Maintenance is this area needs to be considered carefully to minimise risks to apparatus and hence patients. Activities should also include checks on the monitoring systems back to a central reporting point that alerts both clinical staff and maintenance staff to an issue with the medical isolation transformers.

Another more recent development of transformer-based technology has been the use of voltage optimisation. These are effectively 1:1 LV/LV transformers that provide a set voltage output, on the premise that the resultant secondary output voltage is closer to the optimum voltage required by many of the devices connected to an electrical installation. The purpose of the units is to save energy. For resilience and maintenance purposes it would be advantageous for these units to be installed with a system bypass so that the unit can be isolated if required.

HV transformers come in various shapes and sizes. For most user applications a voltage ratio of 11 kV/400 V is used. The main types of construction of the transformers are oil-filled and cast resin. Oil-filled uses different grades of insulating oil. The oil circulates from the warm windings to the transformer fins where it cools by natural convection and returns to the windings again. With a cast resin transformer a fan may be used to drive air past the windings.

With an oil transformer regular checks should be made to ensure that:

(a) the gaskets show no signs of leakage;
(b) the moisture sensor shows no signs of water contamination of the oil; and
(c) the fins and transformer body is showing no signs of corrosion.

Periodically a sample of the oil should be tested and also replaced in accordance with the manufacturer's recommendations. Care should be taken when handling mineral insulating oil. The following standard will provide guidance in this area:

● BS EN 60422:2013 *Mineral insulating oils in electrical equipment. Supervision and maintenance guidance*

For cast resin transformers it is necessary to check:

(a) the operation of any cooling fans;
(b) the operation of any temperature monitoring devices;
(c) for accumulations of dust and debris near the windings or terminals (the transformer may need to be taken out of service for this); and
(d) for condensation near the windings (if the transformer has been out of service for a while).

On all transformers periodic checks (or after a fault event) should be made on the transformer to ensure that:

(a) there is satisfactory insulation resistance to earth on each winding; and
(b) the terminations are mechanically secure.

B1.3 Generators and standby supplies

Principle activities are shown in Table B1.2

▼ **Table B1.2** Generators and standby supplies activity

	Activity	Frequency
	Generator Availability and Servicing	
1	General Fuel level, oil level and radiators, and coolant level should be checked regularly, topped up as required and any remedial activities recorded.	Weekly
2	General Check that automatic transfer switches, generators, fuel pumps etc. are set to 'auto' mode.	Weekly
3	General Check that emergency stop buttons are not activated so that the control system is not inhibited, preventing the generator from an auto start in the event of a mains power outage.	Weekly
4	General Check generator batteries to the manufacturer's recommendations and replace as necessary.	Monthly
5	General The fuel quality should be checked regularly. *If diesel is stored unused for long periods there is potential for some algae and micro-organisms to breed in the fuel.*	Annually
6	General Any major service should be carried out via a 'Permit to Work' system to isolate the generator from standby mode.	Annually
7	Mechanical general Any major service should include the following typical activities, which is a guide only. Further recommendations should be obtained from the manufacturer. Replacement of: • engine oil, with the old oil being disposed of correctly off site, through authorised facilities; • filters including oil, water and fuel; and • air filters as necessary after they have been inspected and cleaned. Check: • wear and tear on all drive belts with replacement as necessary; • wear and tear on all hoses with replacement as necessary; • engine governor and fuel solenoid rates - lubricate as necessary; • engine AV mounting bolts; • fuel pipe work and fuel tank; • condition of the exhaust system; • condition of air intake; and • condition of sound attenuators.	Annually

	Activity	Frequency
8	**Electrical general**	Annually
	Any major service should include the following typical activities, which is a guide only. Further recommendations should be obtained from the manufacturer.	
	Clean the battery terminals and protect with petroleum jelly.	
	Check:	
	• levels of electrolyte in the batteries and top up with demineralised water as necessary;	
	• battery charger for correct operation and its charging rates;	
	• engine heater; and	
	• condition of the generator control panel including, where possible, the connections, contacts and relays.	
	Test and record:	
	• alternator phase to earth insulation resistance;	
	• alternator earth continuity;	
	• the operation of any local fire alarm detection; and	
	• the operation of any local emergency lighting.	
9	**Safety**	Annually
	Check the operation of all generator EPO buttons and associated shut down devices.	
	Visually check all fusible links and associated interfaces to the fuel system and fire alarm system.	
	Generator Testing	
10	Functional testing on building load.	Monthly
	The generator should be tested periodically on load by failing the normal building power supply – this should preferably be simulated by means of a manual switch or built-in key switch on the switchboard. An alternative may be to remove a fuse from the phase failure relay connections.	
	This will test all of the ancillary systems such as the automatic changeover switchgear, the fuel systems and pumps, the generator cooling etc.	
	Run for a minimum of two hours on building load at 70 % minimum alternator capacity (maximum of four hours).	
	Running the generator off load is potentially detrimental to the generator and not recommended.	
11	Functional testing on building load.	Monthly
	All test results shall be recorded and all defects reported.	
	In particular full records should be made at half hourly intervals to assess load profiles (kW, kVA, kVAr, A, V, pf, frequency) and performance criteria including temperature of generators, oil level, water levels, oil pressure etc.	
12	Full capacity testing on 100 % load.	Annually
	The generator should be tested on full load (continuous load rating) – this should be simulated against a fully rated load (load bank) for at least four hours.	
	Allowance should also be made to test the generator at 110 % (standby load rating) for one hour maximum at the end of the main test run.	
13	Full capacity testing on 100 % load.	Annually
	All test results shall be recorded and all defects reported.	
	In particular full records should be made at half hourly intervals to assess load profiles (kW, kVA, kVAr, A, V, pf, frequency) and performance criteria including temperature of generators, oil level, water levels, oil pressure etc.	

B1.4 UPS

Principle activities are shown in Table B1.3

▼ **Table B1.3** Uninterruptable power supply activity

	Activity	Frequency
	UPS Testing	
1	General Any major service should be carried out via a 'Permit to Work' system to isolate the UPS. Switching strategy required to place any UPS unit in system bypass mode to allow servicing.	Annually
2	Operation Check all meters and signal lamps function correctly and replace as required.	Annually
3	Thermographic Survey the UPS connections with calibrated thermal imaging equipment and check for potential hot spots. Note shall be made that a risk assessment will be required for the temporary removal of safety barriers and/or bypassing of safety measures to gain access to switches. Refer to Guidance Note 3 (section 3.11).	Annually
4	For UPS isolations with an external (system) bypass: • transfer UPS to external bypass; • check correct transfer to mains; • switch off UPS output breaker; • power down UPS to isolate from all sources of supply (to include mains input, static bypass, battery and UPS output); and • check system is isolated following the manufacturer's guidance.	Annually
5	For UPS isolations with an internal (system) bypass: • transfer UPS to internal maintenance bypass; • check the correct transfer to mains; • switch off the UPS output breaker of the system; • power down UPS to isolate from all sources of supply (to include mains input, static bypass, battery and UPS output); • allow the d.c. capacitors to self-discharge for recommended time period; and • using appropriate meters, check that internal parts are isolated and proved dead. **Caution:** *restrict access to UPS input and output terminals which will be live.*	Annually
6	General • check UPS batteries in accordance with the manufacturer's recommendations and replace as necessary; and • check UPS and battery connections and make good as required.	Annually
7	Servicing General servicing of the UPS control panels, switchgear and batteries in accordance with the manufacturer's recommendations.	Annually
8	Housekeeping Switchroom not to be used as a store and it should be left in a clean and tidy state. All explosive or flammable materials should be elsewhere in the appropriate enclosures.	Quarterly
9	Operation Check all BMS connections function correctly and repair as required.	Monthly
10	Environment Check the operational environment for abnormal conditions.	Monthly
11	Planned replacement Batteries to be replaced under planned sequence in accordance with the manufacturer's recommendations.	Manufacturer recommendation

B1.5 CHP electrical components

Combined heat and power (CHP) plants are becoming increasingly prevalent. Whilst CHP plants are not a new idea, they are more efficient than in the past and are now a regular feature on larger estates. In simple terms they have an energy input (perhaps gas, biomass or oil) that drives the plant that in turn creates both mechanical heat energy and electrical energy in one single process.

It is beyond the remit of this Guide to go into the full maintenance requirements of CHP plants in too much detail. The electrical component is essentially an electrical generator and associated mains switchgear. Maintenance of this is largely covered in previous sections. A safe system of work must be employed at all times.

Clearly, both the mechanical output and the electrical output of the CHP plant mean that a lot of heat is produced. This will have a detrimental effect on the plant over a sustained period of time. In order to mitigate this deterioration an intensive maintenance regime must be set in place to ensure that the CHP plant is operating at its optimum level for as long as possible.

The Chartered Institution of Building Services Engineers produces a series of Applications Manuals; in particular, AM12 covers combined heat and power for buildings: http://www.cibse.org/knowledge/cibse-am/am12-combined-heat-and-power-for-buildings-(chp)

It is recommended that organisations thinking of using CHP plant should obtain a copy of this document to ensure that the design meets their particular needs. Most of the document focuses on mechanical engineering topics, however, Section 12.3.3 states that:

It is possible for an otherwise well designed and installed CHP plant to be out of action for 1000 hours in a year if faults are not promptly attended to. This may include downtime at times of highest electricity price and greatest heat demand.

A programme of maintenance on CHP plants must also include contingency, or alternative, supply for existing electrical loads that need to remain in service whilst the CHP is offline and undergoing maintenance.

Another consideration is with respect to embedded generation systems, such as CHP plants, is compliance with the requirements of Engineering Recommendation G59/3 from the Energy Network Association and the application of short term operating reserve (STOR). If any form of generator device is connected to run 'in parallel' or 'synchronised' with the mains electrical utility grid (national grid), the requirements of G59 applies to the installation.

A STOR provider must be able to fulfil certain criteria, laid down by the UK national grid, to essentially offset demand on the wider electrical network by using their own embedded generation capacity. This is only really applicable to much larger sites capable of offering a minimum of 3 MW or more of generation or steady demand reduction (this can be from more than one site or more than one generating set). There are stipulations in terms of minimum run times, response times and recovery times.

A CHP plant may be part of this arrangement. Any downtime must be carefully planned and coordinated with the local electricity network otherwise they will be calling on the plant when it is not available. This could have serious issues for resilience across a wider area than just the local site.

It should be noted that the operations and maintenance of CHP plants can be complex and demanding for in-house resources. Some CHP plant suppliers offer an ongoing service support contract, which remotely monitors the performance of the CHP plant and carries out both preventative and reactive maintenance.

B1.6 Solar photovoltaic installations

Installations of solar photovoltaic panels are often marketed as offering relatively low maintenance but, contrary to many installers' sales pitches, they are not 'no-maintenance systems'.

If a system is installed with multiple panels and a single system inverter, then if any cell on any one panel is shaded, perhaps by bird droppings or autumnal leaves, the current flow of the whole system can be affected. So, while the electrical systems of the panels themselves may require very little maintenance, if they are installed in locations that are subject to dirt and dust they will need periodic cleaning to operate at maximum output.

Installation of panel monitoring systems on a single inverter system may assist with the prevention of one panel's deterioration affecting the full operating output of other panels on the system. On more advanced systems monitoring of the performance of each panel's cells is possible.

Regular reviews of their locations with respect to adjacent building projects and shadows may also be necessary. For roof-mounted panels periodic reviews of the fixings with respect to corrosion will definitely be necessary. It is also advisable, on large roof-mounted arrays of solar panels, to have a contingency plan in place for the replacement of a panel.

Water ingress to professionally installed solar panels is unlikely but checks to the inverters and other system equipment is advisable.

Maintenance carried out twice a year should be sufficient to monitor and mitigate any damage before deterioration of the system's performance. Appropriately trained personnel should be used and, with the location often at high level, risk assessments with respect to the Working at Height regulations should be undertaken first.

Principle activities are shown in Table B1.4

▼ Table B1.4 Solar photovoltaic installations

	Activity	Frequency
1	Safety: • assess whether any isolation of the complete system from the mains integration connection is required before any inspection. • a visual inspection and basic cleaning of the panels may not necessitate isolation. • any testing will require isolation of the system and controlled reinstatement. • working at height to be considered at all times.	
2	Solar panels: • debris and dust to be brushed clear or washed off. • inspect for surface damage, cracks or discolouration. • check cable connections for wear and tear or damage. • check panel mounts for corrosion and integrity of the fixings. • check that the fixings maintain the integrity of the roof with respect to water ingress.	Recommend 6 monthly intervals
3	Electrical system check (by appropriately trained personnel only): • check the inverter for signs of water ingress, moisture or rust. • check the inverter for signs of overheating or burning. • check the cable connections are satisfactory. • check that earthing and bonding connections are satisfactory. • check the meter readings, record and compare with previous logs. • check the input and output currents with a d.c. ammeter. • check the input and output voltages with a voltmeter. • where d.c. energy storage systems have been installed directly from the PV output, ensure use of a safe system of work and carry out periodic inspections on batteries etc.	Recommend annually

B1.7 Earthing

The purpose of circuit protective conductors is to provide an electrical system with a controlled safe (low impedance) path for any fault current that may occur in an exposed-conductive-part to flow back to the main body of Earth.

By providing an adequately sized circuit protective conductor, sound mechanical connections for the conductors and a properly rated circuit protective device, a circuit will automatically isolate itself in the event of a fault. In addition, electrical installations will also use protective bonding conductors, which ensure that all extraneous conductive parts are at the same potential as the system earth at the point of supply.

Larger buildings, and some specialist installations, will also have a lightning protection system (see section (h)). Functional earthing is another system that is used in specialist locations, for instance, isolation transformers for specialist care in hospitals and also for telecoms/IT installations.

All forms of earthing are electrically tied together by a mains earth terminal (MET). This is typically a series of wall mounted earth bars located in a main switchroom. Links are usually provided to give a common connection and also to separate each form of earthing from the other during testing purposes.

Earthing and bonding connections should be subjected to a periodic test in accordance with the frequency dictated by BS 7671:2008+A3:2015, however, it is recommended that the installation is risk assessed to take into account the local environment. This risk assessment should be ongoing and should be informed by the quarterly inspections advocated in the table below.

If necessary the frequency of testing must be increased and remedial works carried out to counteract any deterioration in the earthing systems.

Testing should also be carried out immediately after an adverse event caused by either a lightning strike to the installation or by a supply issue that may cause a surge protection device to operate.

Some principle activities are shown in Table B1.5.

▼ **Table B1.5** Earthing conductors activity

	Activity	Frequency
1	Safety: • assess whether any isolation (removal of links) of the system is required before any inspection. • working at height to be considered at all times.	
2	Visual inspection: • inspect for surface damage, cracks, discolouration or corrosion. • check cable connections for wear and tear or damage. • check MET mounts for corrosion and integrity of the fixings.	Recommend quarterly intervals
3	Electrical earthing system check (by appropriate trained personnel only): • check the impedance values on previous certificates. • check the integrity of local earth electrodes. • check the integrity of chemical additives that may have been used with local earth electrodes. • check the cable connections are satisfactory. • check that earthing and bonding connections are satisfactory. • measure the impedance values and compare with previous logs. • check the earth leakage current values and compare with previous logs.	In accordance with BS 7671 and Guidance Note 3

B1.8 Lightning protection

The purpose of lightning protection is to reduce the risk of "loss of human life" and "loss of service to the public", as defined in BS EN 62305. The Standard calls for an annual test and inspection, which is often backed up by insurance requirements, especially for vulnerable structures. Periodic maintenance and associated inspections help to ensure that the safety measures provided remain effective when needed.

Whilst an annual maintenance regime is called for, widely acknowledged best practice within the lightning protection industry recommends a test and inspection at 11-month intervals. This provides a body of knowledge of the performance of the lightning protection system over all seasons through a typical year in all weather patterns. Earth electrode resistance values will vary according to the weather conditions, for instance, where the ground could dry out or become saturated and swell (clay soils for instance).

Physical damage and corrosion can reduce the effectiveness of a lightning protection system. New building extensions, or alterations to the building fabric, may affect the performance of a lightning protection system. An actual strike on the building can affect the performance of the installation in the rare, but possible, event of a secondary strike. In all cases the lightning protection system should be reassessed and tested irrespective of when the next periodic assessment is due.

The relevant British Standard for lightning protection was updated in 2008. The maintenance regime also changed as part of this. For any installation installed after 2008, tests carried out in accordance with BS EN 62305:2011 are required. However, lightning protection installations that were installed before 2008 should be tested in accordance with BS 6651:1999.

Lightning protection systems, by their very design, are typically installed on taller buildings. Regular visual inspections will need risk assessments to consider the implications of the Working at Height Regulations. Specialist companies will have properly trained and equipped staff and so it may be expedient to outsource this particular maintenance task under an appropriate procurement contract.

Annex E.7 of BS EN 62305-3 provides guidance on maintenance and inspection of lightning protection systems.

Clauses 31 to 34 of BS 6651 provide guidance on inspection, testing and maintenance that may be applied to installations that pre-date the publication of BS EN 62305.

B1.9 Surge protection

Both BS 7671:2008+A3:2015 and BS EN 62305 describe the use of surge protection devices (SPDs) at various points within an installation.

The use of SPDs complements and enhances the lightning protection system and should protect the installation against the possibility of damage from direct and indirect lightning strikes. They can also protect against the possibility of voltage transients generated by faulty equipment within an installation, which may in turn affect other vulnerable equipment.

Regular visual checks of the fault indicators on such devices should be undertaken.

During periodic inspections of the main electrical installations steps may be needed to isolate the devices so that they are not damaged by the testing regime, especially during insulation resistance testing. The manufacturer's instructions should be consulted prior to any tests being carried out.

If isolation of the device is not feasible because of electrical safety concerns then a lower voltage (such as 250 V instead of 500 V for single phase installations) for any insulation test may be necessary. Any variation in the testing methodology needs to be noted on the test certificate.

Clause 9.3 of BS EN 62305-4 provides guidance on maintenance and inspection of surge prevention measures.

B1.10 Power factor correction

Power factor correction (PFC) has been used for decades to improve the electrical load characteristics at the point of supply and assist the user to avoid paying penalty charges on a load that is highly inductive. PFC is a useful energy saving system and therefore requires regular maintenance as it will reduce energy bills when operating at its optimum level.

In a modem electrical installation PFC faces a number of challenges to continue to operate successfully:

(a) PFC operating temperature control;
(b) heat gain from adjacent equipment and operating environment;
(c) dust ingress;
(d) humidity; and
(e) electrical infrastructure load characteristics.

Within all electrical installations increasingly sophisticated electronic devices are adding to the harmonic loads of the installation; this has an adverse effect on the operating characteristics of the PFC and can result in electrical resonance on the wider network. Detuned power factor correction capacitors can mitigate this and are usually specified for such applications.

It should also be noted that modern capacitors are designed to fail safe and it may not be immediately apparent that they are not operating satisfactorily.

PFC because of the capacitors, stores energy for a period of time after it has been isolated from the electrical power supply. The ability for the capacitor to retain that energy decays naturally. For this reason only trained staff should access the inside of a PFC cabinet and maintain the PFC system.

A period of time, based on the manufacturer's recommendation, must be allowed to elapse before the cabinet is opened and exposed-conductive-parts are made accessible. Before any work is carried out the equipment must be tested first using a calibrated meter and deemed to be completely safe before touching any terminals.

Some principle activities are shown in Table B1.6.

▼ **Table B1.6** Power factor correction activity

	Activity	Frequency
1	Safety: • isolation and lock off of the system is required before any inspection. • allow any stored energy in the capacitors sufficient time to decay before accessing the inside of the PFC cabinet.	
2	Visual Inspection: • inspect for surface damage, cracks, discolouration or corrosion on the capacitors. • check cable connections for wear and tear or damage. • clean and vacuum all connections, equipment and associated parts including fan filters – removal of dust is vital to prevent flashovers between conductors. • check the capacitor contactors for pitting and general wear and tear – contactors are subject to high inrush currents. • check the operation of any cooling fans in the cabinet and the ventilation of the capacitors. • check the earthing of the capacitors casings.	Recomend annual intervals
3	Electrical operation check (by appropriately trained personnel only and with RAMS in place): • check the integrity of capacitors. • with the system re-energised, the RMS current values can be checked for each capacitor step. • analysis of any harmonics can also be made. • compare results with previous logs. • check the earth leakage current values and compare with previous logs.	Recommend annual intervals (specialist involvement only)

B1.11 Harmonic filters

Harmonic pollution is caused by electronic equipment and their associated power supply units. The harmonic waveform manifests itself as a distortion of the normal sine wave. Variable speed drives, in particular, whilst employed to provide soft starting of heavy motor loads and energy efficient running, also cause a number of problems.

Harmonics increase the RMS value and the peak value of the current waveform, which then causes increased heat in electrical equipment. Circuit breakers could operate because of higher thermal or instantaneous levels.

Other adverse effects may include:

(a) damage to capacitors;

(b) blown fuses;

(c) false readings on KWh meters;

(d) damage to sensitive electronic equipment;

(e) electronic communications equipment interference;

(f) synchronisation issues;

(g) excessive neutral currents;

(h) overheating of cables;

(i) overheating of transformers; and

(j) premature ageing of equipment.

Use of active harmonic filters can reduce these effects. ENA Engineering Recommendation G5/4-1 (Oct 2005) provides guidance on acceptable levels of harmonics within an installation.

It is the installations user's responsibility to ensure that their load does not exceed those levels as it will affect the wider network too. Where it is proved that an installation load is persistently causing too much harmonic distortion to the wider network an electricity supplier may, as a last resort, take steps to disconnect the supply.

Assessing an installation's harmonic profile and selecting the correct level of active and passive filters is a specialist task. As a maintainer, a regular assessment of the load characteristics of the equipment connected to the building would be a first step, which will reduce a potentially expensive survey fee.

Keeping an asset register up to date has many benefits across the whole of the site – including the harmonic details of any new and existing equipment could add to that body of knowledge.

Some principle activities are shown in Table B1.7.

▼ **Table B1.7** Harmonic filters activity

	Activity	Frequency
1	Safety: • isolation and lock off of the system is required before any inspection. • allow any stored energy in the capacitors sufficient time to decay before accessing the inside of the harmonic cabinet.	
2	Visual inspection: • inspect for surface damage, cracks, discolouration or corrosion on the capacitors. • check cable connections for wear and tear or damage. • clean and vacuum all connections, equipment and associated parts including fan filters – removal of dust is vital to prevent flashovers between conductors. • check any contacts for pitting and general wear and tear. • check the operation of any cooling fans in the cabinet and the ventilation of the capacitors. • check the earthing of the casings.	Recommend annual intervals
3	Electrical operation check (by appropriately trained personnel only and with RAMS in place): • check the integrity of the filters. • analysis of any residual harmonics can also be made. • compare results with previous logs.	Recommend annual intervals (specialist involvement only)

Appendix B2 General applications and circuit protection

Introduction

Whilst there are undoubtedly many specialist installations that require satisfactory electrical maintenance in varying environments and industries, by far the bulk of maintenance activity will be run-of-the-mill small power and lighting installations.

For small power sub-circuits the central documents to follow will be BS 7671:2008+A3:2015 and the associated Guidance Note 3: *Inspection & Testing*. These will provide definitive guidance on periodic testing and inspection of existing installations. Guidance Note 7: *Special Locations* will also provide information.

Whilst many smaller installations will be able to conform to one complete periodic test and inspection every few years in accordance with the guidance of BS 7671:2008+A3:2015, larger installations can be too big to consider completing all in one go. In larger estates the realities are that whole departments are moved every so often as their needs change. Within larger hospitals the concept of decant wards is used so that a spare ward can be refurbished before everyone shuffles around.

Testing and inspection in these examples could be split into more manageable chunks of work so that, over an agreed period of time, all of the estate is covered. Of course if an area is refurbished then the opportunity for test and inspection is there whilst the premises are vacant.

With general lighting installations a more frequent maintenance regime may be necessary and will probably be based on a number of factors as discussed below.

Standards

- BS 7671:2008+A3:2015
- Guidance Note 3: *Inspection & Testing, 7th Edition*
- BS EN 12464-1:2011 *Light and lighting. Lighting of work places. Indoor work places*
- BS EN 12464-2:2014 *Light and lighting. Lighting of work places. Outdoor work places*

B2.1 Protective devices, overloads, discrimination and grading

Electrical installations, especially in larger estates, will evolve over time as the purpose of a building alters and adapts to new requirements. Loads are added either by design or in iterative steps.

At the original design stage it is reasonable to assume that the infrastructure was designed to accept future loads. However, it is prudent to periodically check that all protective devices are correctly rated for their existing load as it is now.

On adjustable protective devices it may be feasible to update the settings to a new load, however, it should be double checked that they still successfully grade with the upstream devices. If the upstream infrastructure has not changed, despite the additional loads imposed on it, maximum loads and diversity should be re-assessed and the findings documented.

Uncontrolled adding of new electrical loads could be detrimental to the whole installation. All too quickly the 10 % or 25 % spare capacity that may have been allowed for in the original design will no longer be there.

Some principle activities are shown in Table B2.1.

▼ **Table B2.1**   Protective devices, overloads, discrimination and grading activity

	Activity	**Frequency**
1	General	5 yearly
	Audit of the original design load criteria of the protective devices to be checked against the current loads and updated protection study carried out, if required, in accordance with BS 7671:2008+A3:2015.	
	Any necessary adjustments to switchboard protective devices are to be recorded.	

B2.2 General lighting

General lighting is provided to illuminate areas that cannot be satisfactorily lit purely by natural daylight or to allow activities to take place after dusk. As such, lighting designs for different areas provide for different levels of light. Walking around a car park will not require the same levels of light as working at a desk.

All designs allow for a lighting maintenance factor – this is based on a number of criteria including the environment in which the luminaires are installed and accessibility for lamp changes. This takes into account cleanliness of lamp optics and covers to ensure that light levels are satisfactory for as long as possible.

Maintenance factors for luminaires will vary according to the lamp source type, the environment in which they are installed and also the intervals between cleaning. Measured on a scale of 0 to 1, typical values can be as low as 0.45 in harsh environments with infrequent cleaning or as high as 0.95 in benign areas with twice yearly cleaning.

A design with a maintenance factor of 0.8 will be lit to 125 % of the requirement at the moment of commissioning.

Lifespans of lamps within luminaires can also vary, again according to the lamp source and the environment in which they are installed. Other influences such as the pattern of use, lighting controls and the quality of manufacture will also have an effect. This is all summarised as a lamp survival factor.

It is incumbent for maintenance teams to react to reports of lamp failure as soon as possible, especially in areas where there is a large amount of pedestrian and vehicular traffic.

Lamp failure can cause considerable visual discomfort and safety concerns for occupants of the space. Within car parks, for instance, concerns for personal safety after dark will be increased. Lamp failures in offices may lead to lower productivity and, within factories, safety risks with moving machinery.

Going forward lighting installations of many kinds will increasingly be using LED-based solutions. LED lighting installations, whilst being perfectly viable choices, still need to be carefully considered. The following publication will assist with this:

IET *Code of Practice for the Application of LED Lighting Systems* (2013)
http://www.theiet.org/resources/standards/led-cop.cfm

When undertaking maintenance for luminaires a number of safety concerns need to be borne in mind:

Some principle activities are shown in Table B2.2.

▼ Table B2.2   General lighting activity

Item	Element	Maintenance Issue	Control measure to be considered
1	Safe working	Invariably, luminaires are located on ceilings or at a high level on walls or in staircases.	Appropriate platforms or stepladders for safe access which themselves have also been tested and certified.
2	Safe working	For long linear luminaires (1.8 m/2.4 m) at very high levels, access to both ends is difficult.	Appropriate mobile platforms covering the whole length of the luminaire for safe access plus two members of maintenance staff.
3	Safe working	Pedestrian traffic in area below luminaires.	Appropriate diversions or barriers to separate public from works area.
4	Safe working	Pedestrian traffic in area below luminaires.	Consider out of hours work – possibly more expensive resources but quicker activity with less control measures for non-maintenance occupants required – restrict access to whole building except for maintenance staff.
5	Safe working	Existing luminaires isolated for maintenance purposes.	Temporary lighting may be required.
6	Electrical isolations	Luminaire fixtures and terminals may be live even if lamps have failed.	Ensure circuits are isolated at switch as a minimum during lamp changes, only where there is clear line of sight from the luminaire to the switch.
7	Electrical isolations	Luminaire fixtures may be live even if lamps have failed and have two-way switching on the circuits.	Ensure circuits are isolated at distribution board as a minimum safety standard during lamp changes complete with warning notices.
8	LED retrofit	Luminaires may have been adapted with control gear bypassed when linear LED retrofit lamps were first installed – one terminal will be a 0 V potential and the other end at 230 V.	Ensure that: • luminaire is isolated; • terminals are clearly identified; and • replacement LED lamp is installed the correct way around.
9	Lamp changes	Traditional lamp sources are typically contained within glass enclosures. Risks of lamps breaking when being changed.	Handle with care and wear protective gloves to avoid cuts and abrasions.
10	Fire integrity	Recessed luminaires in fire rated ceilings should have fire rated covers to maintain the integrity of the ceiling.	Check covers are still in place and still in good state of repair or replace.

B2.3 Small power

The tables below are practical guidance based on the example of a large data centre type installation that could be difficult to test and inspect in one go once every five years. By testing specific parts of the installation (circa 20 %) the entire installation will be tested over a period of five years. Some non-critical areas can be tested in their entirety in one go.

Another strategy could be to test 25 % of the critical parts of the installation every year and use the fifth year to catch up with remedial works. This may help to ease pressure on resources and maintenance budgets.

Whatever strategy is employed it is suggested that this is documented, perhaps as part of a site electrical safety and maintenance policy document. This can then be corroborated as part of any external audit and used to demonstrate that the regime fulfils the statutory requirements.

Some principle activities are shown in Table B2.3.

▼ **Table B2.3**   Small power activity

	Activity	Frequency
1	**General** Functional inspection and testing carried out in accordance with BS 7671:2008+A3:2015 and Guidance Note 3 (section 3.10.2 & table 3.3) and serviced in accordance with the associated manufacturer's recommendations.	5 yearly
2	**General** Plan and record the number of circuits to be tested – not to be less than 20 % annually within the data hall on a rolling programme to achieve 5 year plan overall. Other non-critical areas to be tested area by area, again on a rolling programme.	20 % annually
3	**Identification** Identify circuits to be tested and remove all electrical loads from socket outlets – where possible place on alternative supplies.	20 % annually
4	**Testing** Testing of main distribution circuits from transformers to switchboards. To include earth continuity, polarity check, insulation resistance, earth loop impedance and prospective short circuit current.	5 yearly
5	**Testing** Testing of sub-distribution circuits from switchboards through UPS to PDUs. To include earth continuity, polarity check, insulation resistance, earth loop impedance and prospective short circuit current.	20 % annually
6	**Testing** Testing of sub circuits from PDUs to cabinets. To include earth continuity, polarity check, insulation resistance, earth loop impedance and prospective short circuit current.	20 % annually
7	**Testing** Testing of sub-circuits from switchboards to primary plant. To include earth continuity, polarity check, insulation resistance, earth loop impedance and prospective short circuit current.	20 % annually
8	**Testing** Testing of sub-circuits from switchboards to non-essential equipment and loads. To include earth continuity, polarity check, insulation resistance, earth loop impedance and prospective short circuit current.	5 yearly
9	**Testing** All test results shall be recorded and all defects reported.	20 % annually

Electrical distribution RCDs

In recent years, following developments within BS 7671:2008+A3:2015, there have been increasing numbers of RCDs installed either as part of a faceplate (13 A socket or fused spur) or as part of a protective device (RCBO) at the distribution board. These need checking periodically for function and also time response.

Some principle activities are shown in Table B2.4.

▼ **Table B2.4** Electrical distribution RCD's activity

	Activity	Frequency
1	General Functional inspection and testing be carried out in accordance of BS 7671:2008 +A3:2015, Guidance Note 3 and the manufacturer's recommendations.	Quarterly
2	Identification Identify circuits to be tested and remove electrical loads from socket outlets.	Quarterly
3	Testing Check the functional operation of RCD using the built-in test button to the requirements of BS 7671:2008+A3:2015. All test results shall be recorded and all defects reported.	Quarterly
4	Testing Check the operation of the RCD using calibrated proprietary test equipment to the requirements of BS 7671:2008+A3:2015. All test results shall be recorded and all defects reported.	Annually

B2.4 Electrical heating of hot water

Engineered water systems within the built environment are one of the greatest advances in human civilisation. However, it must be recognised that poorly designed and badly maintained water systems can also create as many problems as they solve. Contaminated water poses huge health risks, especially to those vulnerable with low immunity to infections. Maintenance teams must ensure that they have the correct level of skills and technical knowledge on hot and cold water systems and other engineered water systems that might cause a risk to human health.

An organisation must put into place control measures and risk management strategies to implement, manage and review engineered water systems, at regular intervals, to reduce the risk and effects of bacterial infections such as legionella and pseudomonas.

For larger installations the use of electricity to heat water is not common because of the costs involved in using large amounts of energy from electrical sources. It is more common to use gas fired boilers or similar mechanical engineering installations because the inherent energy costs are usually lower. However, for smaller installations, or off gas grid installations, it may be more expedient to use electricity to heat water.

It is beyond the remit of this particular Guide to provide an in-depth analysis of the risks of legionella and associated issues. The following is a quote from the BSI webpage providing information on BS 8580:2010:

> *It is the responsibility of the duty holder to ensure that an assessment is carried out to identify and assess the risk of exposure to Legionella from work activities and water systems and to put in place any necessary precautions*

There is a great deal of industry information freely available from a number of resources and the reader should conduct their own research into this using some of the following leads, which is not an exhaustive list:

(a) BS 8580:2010 *Water quality. Risk assessments for Legionella control. Code of practice*
(b) Health and Safety Executive website information on Legionella
http://www.hse.gov.uk/legionnaires/
(c) HSE L8 ACOP 4th Edition 2013 *Legionnaires' disease: The control of legionella bacteria in water systems*
(d) HSE HSG 274 *Legionnaires' disease; Technical guidance*
(e) CIBSE Technical Memoranda *TM13 - Minimising the Risk of Legionnaires' Disease 2013*
(f) Health Technical Memorandum 04-01 *The control of Legionella, hygiene, "safe" hot water, cold water and drinking water systems*

   Part A: Design, installation and testing
   Part B: Operational management

(g) BSRIA Legionnaires' Disease – Risk Assessment (BG 57/2015)
(h) BSRIA Legionnaire's Disease – Operation and Maintenance Log Book (BG 58/2015)

For an electrical systems maintainer it is incumbent to conduct regular checks on their electrical water heating systems, with particular emphasis on the following:

Some principle activities are shown in Table B2.5.

▼ Table B2.5 Electrical heating of hot water activity

	Activity	Frequency
1	Maintain a schedule of all water heaters within an installation.	Quarterly
2	Maintain a schedule of all sensor taps within an installation.	Quarterly
3	Check the outgoing water temperature from the hot water cylinder (should be at least 60 °C).	Monthly
4	Where applicable, check the return water temperature to the hot water tanks (should be at least 50 °C).	Monthly
5	Check the temperatures at the hot water sentinel outlets (should reach at least 50 °C within a minute of running the outlet). NOTE: *if a thermostatic mixing valve (TMV) is present, use a surface temperature probe at the inlet pipe to the TMV.*	Monthly
6	Check the temperatures at the cold water sentinel outlets for each cold water tank (should be less than 20 °C within two minutes of running the outlet).	Monthly
7	Record which hot water outlets are in areas not frequently used.	Fortnightly
8	Reduce the risks of legionella by conducting regular flush-through exercises on systems not regularly in use.	Weekly
9	Manage adaptations to the water pipework to reduce dead legs.	Quarterly
10	Conduct a regular purge on hot water storage systems by providing a boost to raise the temperature above 60 °C.	Weekly

A local risk assessment will also add to this basic list. Additionally, a review of the operating conditions and frequency of use will also inform the frequency of checks and testing on the water systems; for instance, have areas of a building been mothballed or recently vacated?

Newer technologies such as air source and ground source heat pumps and also refrigerant based thermodynamic heater panels are increasingly being used for heating water. Whilst being designed to reduce the carbon based energy needed to heat water by harvesting energy from ambient temperatures, these systems typically raise the temperature of the water to levels below 60 °C. Such systems therefore need close monitoring and regular boosts to reduce legionella risks.

B2.5 d.c. systems

With increasing numbers of electronic equipment being used to assist us with our daily lives, both at work and at home, the rationalisation of energy supplies to provide power for these devices is inevitable. To date all these devices have come with a.c. adapters and rectifiers. d.c. electrical micro generation devices, including photovoltaic panels, are also installed with inverters to connect to legacy a.c. distribution systems.

Increasingly, with a drive towards energy efficiency, the trend will be towards complete d.c. systems from energy intake to point of use. Demonstrator installations are already in place in the UK, for example:

(a) a London theatre with a PV array connected to battery systems that provide power for LED lamps; and
(b) power over ethernet installation near London Bridge providing power modular lighting in a meeting room.

The IET *Code of Practice Low and Extra Low Voltage Direct Current Power Distribution in Buildings* (2015) discusses the types of d.c. systems already being developed and the parameters that will make them safe to design and install. It includes new and distinct d.c. installations with a central inverter near the traditional a.c. distribution board and a separate d.c distribution system closer to the point of use to overcome volt drops. There are also sections on the conversion of existing telecoms and data networks to accept small electrical loads. Clearly, the issue here is heat gain on cables not originally designed for this in bundles on a cable tray with no means of heat dissipation. Other installations discussed within the Code of Practice are the conversion of legacy a.c. systems to provide separate d.c. infrastructure within buildings etc.

In terms of maintenance, the same general principles apply as for a.c. systems:

(a) isolate and make safe;
(b) maintain;
(c) test and inspect; and
(d) reinstate.

Ultimately, it is important to remember that an electrical fault current on a d.c. system can still be lethal under certain circumstances. Safety is still paramount and risks need to be managed in the same way.

Appendix B3 Life safety systems

Introduction

Within a building, life safety systems provide an automated means of alerting occupants to the dangers of fire, smoke or other hazards, such as gas. To aid evacuation in such circumstances a public building must be provided with emergency lighting. The main design standards within the UK are listed below. Maintaining such systems to ensure they operate satisfactorily when required is a statutory requirement and is clearly a major obligation for the maintenance team.

Standards

- BS 5266-1:2011 *Emergency lighting – Part 1: Code of practice for the emergency escape lighting of premises*

- BS EN 50172:2004, BS 5266-8:2004 *Emergency escape lighting systems*

- BS 5839-1:2013 *Fire detection and fire alarm systems for buildings. Code of practice for design, installation, commissioning and maintenance of systems in non-domestic premises*

- BIP 2109:2013 *The design, installation, commissioning and maintenance of fire detection and fire alarm systems in non-domestic premises. A guide to BS 5839-1:2013*

B3.1 Emergency lighting

Clause 7.2 of BS 5266-8 covers the routine inspection and testing of emergency lighting systems and defines a number of daily, monthly and annual activities. The annual testing is further reinforced by the issuing of certificates.

Further information on the maintenance of emergency lighting systems can also be found in the IET publication *Electrician's Guide to Emergency Lighting 2nd Edition* (2014), which provides examples of:

(a) installation commissioning certificates;
(b) periodic test and inspection certificates;
(c) inspection and test records; and
(d) fault action records.

Emergency luminaires should be clearly identified on layout drawings and ideally should each have a unique reference number. This allows a schedule of tests to be collated over the course of a year and also to ease maintenance and replacements where necessary.

Some principle activities are shown in Table B3.1.

▼ **Table B3.1** Emergency lighting activity

	Activity	Frequency
1	System status: • visually check that all maintained lamps are operating via the indicator lamps. • visually check that all system healthy indicators on any central battery systems are illuminated. • check that any system fault is recorded and given urgent attention. • record activities in a logbook, note all remedial works that are required and escalate items for further attention.	Daily
2	Short duration test: • check all luminaries and other emergency lighting equipment is in a good condition, all lamps and light controllers are clean, undamaged and not blackened. • briefly test all emergency lighting equipment by simulating a failure of the normal lighting supply. The test should not exceed a quarter of the equipment rated duration. Check that all equipment functions correctly. • check that, upon restoring the mains supply, all supply healthy indicators are again illuminated. • record activities in a logbook, note all remedial works that are required and escalate items for further attention.	Monthly
3	Full duration test: • a full system test should be conducted by a competent service engineer, including a full rated duration test of the system. • compliance of the installation and system with the requirements of BS 5266/BS EN 1838 should be considered and documented. • record activities in a logbook, note all remedial works that are required and escalate for further attention.	Annually

B3.2 Fire detection and alarm systems

BS 5839-1:2013 provides information on the maintenance of fire detection and fire alarm systems in non-domestic premises. It defines a number of weekly, monthly, quarterly, 6-monthly and annual activities. Annual certification is required to prove the status of the system.

Further information on the maintenance of fire detection and alarm systems can also be found in the IET publication *Electrician's Guide to Fire Detection and Alarm Systems* 2nd Edition (2014).

Within BS 5839-1:2013, Section 6 provides further explanation on routine testing, inspection and servicing and non-routine attention, whilst Section 7 gives coverage of the user's responsibilities with details on premises management and the use of a logbook to record all activities.

Routine testing by the user

The purpose of routine tests is to ensure that systematic failure of the entire fire detection and fire alarm system does not go unnoticed.

Some principle activities are shown in Table B3.2.

▼ **Table B3.2** Fire detection and alarm systems routine user testing activity

	Activity	Frequency
1	• Test a manual call point during working hours to check satisfactory operation of the control indicating equipment and the alarm sounders. • Rotate around the installation so that each week a different manual call is tested (note each call point tested in the logbook to provide continuous record for audit purposes). • Test any voice alarm systems each week (refer to BS 5839 Part 8). • For installations such as hospitals it is very important that any systems linked to an alarm receiving centre (ARC) is notified before the start of testing and on completion.	Weekly
2	Other advisable checks (non-BS 5839) include: • are all escape routes clear and is the floor in good condition? • can all fire escapes be opened easily? • do automatic fire doors close correctly? • are fire exit signs in the correct place?	Weekly
3	• Test any standby generator where the set is used to support the fire detection and fire alarm system. • Inspect any vented batteries used as a standby power supply for the fire detection and fire alarm system.	Monthly

Inspection and servicing by a competent person

The purpose of inspection and servicing is to ensure that the fire detection and fire alarm system is fully serviceable and able to operate satisfactorily at a time of need.

Additionally, fire alarm maintenance teams, either in-house or outsourced, should advise their clients of problems with system design, apparatus or detector selection and positioning, which could lead to the system generating false alarms.

However, it should be remembered that, under the terms of BS 5839-1:2013, the required major servicing events carried out by competent persons are not a time to condemn systems that may have complied with previous versions of the Standard but do not quite reach the latest requirements. There is a duty of care to report and inform; then await further instructions. A standard does not normally require retrospective action.

The measures to limit false alarms are divided into eight groups:

(a) siting and selection of manual call points;
(b) selection and siting of automatic fire detectors;
(c) selection of system type;
(d) protection against electromagnetic interference;
(e) performance monitoring of newly commissioned systems;
(f) filtering measures;
(g) system management; and
(h) regular servicing and maintenance.

Periodic inspection and testing

A fire risk assessment (FRA) of the type of the installation and occupancy levels should determine the length of periods between periodic inspection and servicing. The typical recommended period between successive inspection and servicing visits should not exceed six months.

Some principle activities are shown in Table B3.3.

▼ **Table B3.3** Fire detection and alarm systems inspection and testing activity

	Activity	Frequency
1	• Inspect the logbook. • Visual inspection of any structural or occupancy changes and associated impacts on the fire detection and fire alarm system. • Check the records of unwanted fire signals (false alarms) and ensure that the relevant action is taken. • Check and test batteries. • Check and test control panel functions. • Check and test call points and automatic fire detection devices. • Check sounder, bells and visual alarm indicators. • Ensure that all occupants can hear and/or see the alarms and indicators. • Check connections to any ARC. **Caution:** *ensure that the ARC is notified before the start of testing and on completion.* • Check and test all fault indicators and circuits. • Test any output printers. • Carry out any further checks and tests recommended by the manufacturer. • Report any defects. • Update the logbook with any actions completed and issue servicing certificate.	FRA/6 months intervals
2	• Competent person to check vented batteries and top up if necessary.	Quarterly
3	• Test every manual call point. • Examine and functionally test every automatic fire detector including: • smoke detectors; • resettable heat detectors; • optical beam smoke detectors; • aspirating fire detection systems; • carbon monoxide fire detectors; and • flame detectors. Examine and functionally test every fire alarm device, both for visual and audible function. • Where required, replace filament lamps. • Check signal strengths on radio fire detection and fire alarm system. • Visual inspection of readily accessible cable fixings (note that BS 7671:2008+A3:2015 requires that these be fire rated). • Check the cause and effect programme and correlate to any layout changes or zone changes. • Check the standby power supply capacity. • Carry out any other annual checks and tests as recommended by the system component manufacturers. • Carry out visual inspections of motorised fire dampers and check satisfactory operation. • Report any defects. • Update the logbook with any actions completed and issue servicing certificate. **NOTE:** *there is provision within BS 5839 to spread this work over two or more service visits during each twelve-month period rather than over one visit.*	FRA/12 months intervals/2 × 6 months intervals

B3.3 Gas/carbon monoxide detection

Within the domestic and residential sectors, where gas appliances are used for heating and hot water, regular maintenance and testing of the gas installation is extremely important but beyond the remit of this Guide.

Carbon monoxide is an unwelcome by-product of natural gas heating processes. With a well maintained gas system that is properly ventilated there will be no problem. However, if ventilation flues become blocked there is a real risk that carbon monoxide will leak into the occupied areas. Carbon monoxide poisoning is lethal. Detecting leaks from such systems to provide early warning to occupants is very important.

Where standalone carbon monoxide detectors have been provided it is important to regularly check that the batteries are working and that the warning alarm is audible in accordance with the manufacturer's recommendations. If carbon monoxide detectors are linked to a fire alarm system then the manufacturer's recommendations must be followed periodically to ensure that they operate satisfactorily.

A selection of standards to assist in this particular area is as follows:

- BS 7967:2015 *Guide for the use of electronic portable combustion gas analysers for the measurement of carbon monoxide in dwellings and the combustion performance of domestic gas-fired appliances*
- BS EN 50292:2013 *Electrical apparatus for the detection of carbon monoxide in domestic premises, caravans and boats. Guide on the selection, installation, use and maintenance*
- BS EN 50291-1:2010+A1:2012 *Electrical apparatus for the detection of carbon monoxide in domestic premises. Test methods and performance requirements*
- BS EN 50545-1:2011 *Electrical apparatus for the detection and measurement of toxic and combustible gases in car parks and tunnels. General performance requirements and test methods for the detection and measurement of carbon monoxide and nitrogen oxides*
- BS 5446-3:2015 *Detection and alarm devices for dwellings. Specification for fire alarm and carbon monoxide alarm systems for deaf and hard of hearing people*
- BS EN 60079-29-2:2007 *Explosive atmospheres. Gas detectors. Selection, installation, use and maintenance of detectors for flammable gases and oxygen* provides guidance on the selection, installation, safe use and maintenance of electrically operated group II apparatus within explosive atmospheres.

According to the BSI website, BS EN 60079-29-2:2007 applies:

to apparatus, instruments and systems that indicate the presence of a flammable or potentially explosive mixture of gas or vapour with air by using an electrical signal from a gas sensor to produce a meter reading, to activate a visual or audible pre-set alarm or other device, or any combination of these.

BS EN 60079-29-2:2007 is a guide for use and states that daily performance testing, with a small amount of gas, of all gas detectors (known as 'bump testing') is recommended. Again, a risk assessment is advised when creating a maintenance and health and safety policy in this area. It may be considered that daily testing should be mandatory.

B3.4 Personnel alarms and call systems (security, disabled persons, patients)

Disabled toilet alarms

Building regulations in the UK require that alarms are installed within disabled persons' toilets. These pull-cord operated devices are used to alert other occupants of the building if the disabled person needs urgent assistance. The devices should comply with the requirements of BS 8300:2009+A1:2010 *Design of buildings and their approaches to meet the needs of disabled people. Code of practice.*

Periodic checks should include that:

(a) the power supply is functioning correctly;
(b) the alarm is functioning correctly and is both visual and audible in the toilet;
(c) the alarm is functioning correctly and is both visual and audible in a public area (i.e. adjacent corridor) and/or centrally at a manned security point or reception desk;
(d) the resets are functioning correctly; and
(e) no fraying on any cords or damage to any push buttons.

Lift communications systems

For information on the maintenance of lifts in general the reader should refer to BS EN 13015:2001+A1:2008 *Maintenance for lifts and escalators. Rules for maintenance instructions.*

Lifts are required to have alarms and two-way communication devices installed to alert an external agency to the fact that persons are trapped inside a lift. This is especially important out of normal hours when the building may be largely unoccupied. A separate telephone line is usually installed to the lift controller, either in the lift motor room or directly in the lift shaft.

The standard relating to this is BS EN 81-28:2003 *Safety rules for the construction and installation of lifts. Remote alarm on passenger and goods passenger lifts.* It applies to alarm systems for all types of passenger and goods passenger lifts, and deals with hazards including entrapment of users due to the lift not working properly.

Emergency voice communication systems

Any building over one storey in height must have disabled persons' refuge spaces in fire rated areas, typically in or near protected stair wells. These allocated spaces provide a refuge space for up to 30 minutes during the evacuation of a building. The purpose of emergency voice communication (EVC) systems is to allow disabled persons to alert others that they are in the refuge space and need assistance to evacuate the building.

BS 5839-9:2011 *Fire detection and fire alarm systems for buildings. Code of practice for the design, installation, commissioning and maintenance of emergency voice communication systems* defines two types of outstation:

(a) type A can be used as a fire telephone or disabled refuge call point; and
(b) type B outstations can only be used where the background noise is below 40 dBA (therefore there can be no sounder or voice alarm coverage in the area).

Periodic checks should include that:

(a) the power supply is functioning correctly – this should be a dedicated supply using fire rated cable;

(b) the communication system is functioning correctly and is clearly audible at the refuge point;

(c) the communication system is functioning correctly and is clearly audible at a centrally manned security point or reception desk, ideally close to the main fire alarm panel; and

(d) no damage to the fascia of the refuge call point.

Personnel attack alarms

In a number of public locations, such as hospitals, local health centres, retail outlets, banks etc., where staff may be vulnerable to rare attacks from members of the public, patients or other visitors, it is advisable to provide a staff attack system that alerts other staff and security personnel so that assistance can be provided.

Such systems may be hard-wired with call points on reception desks or wall mounted but can also be wireless. Periodic checks should include that:

(a) the power supply is functioning correctly to all parts of the system;

(b) if a wireless system is provided, regular assessment of interference from other wireless systems is made;

(c) if a wireless system is provided, functional checks in all parts of the building are made;

(d) if wireless boosters are deployed, these should be checked for correct operation;

(e) the communication system is functioning correctly and is clearly audible within the local department;

(f) the communication system is functioning correctly and is clearly audible at a centrally manned security point or reception desk;

(g) no damage to the fascia of the staff attack call points;

(h) a sample of static call points is operating correctly – some types have double knock approach or require pressure from two fingers simultaneously; and

(i) all resets are functioning correctly.

Patient call/nurse call systems

Within hospitals there is a requirement, under Healthcare Technical Memorandum (HTM) 08 03 Bedhead Services, to provide a communication system that provides "the ability for patients to summon nursing assistance at the bed space or nursing location". There is also an operational requirement for "clinical staff to communicate remotely with the patient, and with each other". Section 10.2 states that this "is an essential life safety component of bedhead services".

Section 10.2 of HTM 08 03 describes the following integrated systems:

(a) patient-to-nurse (non-speech) calls; and
(b) staff-to-staff (emergency calls).

Nursecall systems also allow system integration from associated services within the clinical areas of a hospital. These may include:

(a) patient equipment – monitoring to link to equipment being used in the care of patients;
(b) bed status – to monitor availability of bed spaces;
(c) bed exit – to monitor patients leaving beds without medical supervision;
(d) fire alarm secondary indication – some areas will only have visual alarm indicators to avoid panicking the less able and evacuation is under medical staff supervision;
(e) medical gas alarm indicators – although there will be system indicators within an estate's office location, links to the local nursecall may be beneficial; and
(f) door entry – security within hospitals is an issue and local interface with the nursecall system at the nursebase screen is beneficial.

Such systems may be hard wired or wireless. Periodic checks should include that:

(a) the power supply is functioning correctly to all parts of the system;
(b) if a wireless system is provided, regular assessment of interference from other wireless systems are made;
(c) if a wireless system is provided, functional checks in all parts of the building are made;
(d) patient-to-nurse system functioning correctly and clearly audible within the ward;
(e) nurse-to-nurse system functioning correctly and clearly audible within the ward;
(f) types of alarm signal are distinguishable from each other and in accordance with HTM 08 03 Table 2;
(g) the follow on lights respond to various scenarios in various locations throughout the ward;
(h) no damage to the fascia of the call points; and
(i) all resets are functioning correctly.

Appendix B4 Industrial and control

Introduction

The following specialist areas will all require attention from appropriately trained professional maintenance staff. It may well be that outsourcing may be expedient and more cost effective.

B4.1 Hazardous areas and equipment

Working in hazardous areas, which may have explosive atmospheres, is a highly specialised activity that should only be carried out by suitably trained professionals using approved procedures, appropriate equipment and PPE. The installed electrical systems within such hazardous areas must all be properly designed. The materials and designated equipment in these areas are designed to reduce any risks of sparks and explosions.

The BSI series of standards BS EN 60079 provides guidance in a number of technical areas with regards to these types of installations. Relevant to this Guide is BS EN 60079-17:2014 *Explosive atmospheres. Electrical installations inspection and maintenance.*

There is a joint publication by the Energy Institute and the Association for Petroleum and Explosives Administration, entitled *Design, construction, modification, maintenance and decommissioning of filling stations* (3rd edition 2011). Various engineering issues, including electrical installations, are discussed, as well as maintenance and decommissioning of filling stations, and gives information to mitigate several risks such as fire, explosion, health and the environment.

The Engineering Equipment & Materials Users Association (EEMUA) Publication 186 *A Practitioner's Handbook for potentially explosive atmospheres* (2014), now in its sixth edition, offers guidance for safe design, installation, inspection and maintenance work in potentially explosive atmospheres.

The IET's Wiring Matters magazine Issue 42 contained an article on filling stations that provides further background information on these issues: http://electrical.theiet.org/wiring-matters/42/.

B4.2 Motors and variable speed drives

Variable speed drives (VSD), whether as new installations or as retro-fitted devices, provide a means to closely control the operations of motors. This close control allows the motors to work more efficiently and they are typically deployed to save energy. However, VSDs will only save energy if they have been specified correctly and they are maintained properly to ensure optimum performance. Maintenance of VSDs is often ignored to the detriment of the wider electrical installation.

VSDs are fundamentally electronic systems with very few moving parts. Their operation produces heat that is removed with heat sinks and cooling fans. The fans use air filters to reduce the risk of dust getting to the electronic components. The fans have a finite lifespan and the filters require periodic replacement.

Loose terminations, like most electrical equipment, can be cause for concern with VSDs. These generate heat and arcing, which can result in failure of the device.

VSDs should be checked regularly for optimum performance:

(a) ambient temperature – circa 25 °C is optimum.
(b) operating parameters including current, voltage, frequency, harmonics etc.
(c) functional checks on the VSD cooling fans.
(d) functional checks on any alarm connections and settings.

Annual maintenance checks should be made, which will necessitate isolations of the VSD and motors supplies. If the VSD is isolated time must be allowed for any integral capacitors to discharge satisfactorily before removing covers. These more intrusive annual checks may include:

(a) vacuuming dust and dirt from heat sink fins;
(b) cleaning or replacing intake air filters;
(c) checking ventilation fans for proper operation and noise;
(d) checking circuit electrical connections for corrosion and loose terminations;
(e) checking CPC connections;
(f) checking thermographic image of the VSD input and output;
(g) checking voltage readings; and
(h) checking the integrity of the VSD enclosure.

B4.3 Control panels

Within plantrooms motor control panels (MCP, also known as motor control cabinets, MCC) are used to marshal all the controls for various items of the plant that are linked to mechanical equipment and systems. Often 400 V or 230 V supplies for a mechanical plant are also routed through MCPs via protective devices. These panels are types of electrical switchboards.

BS EN 61439-2:2011 *Low-voltage switchgear and controlgear assemblies* describes forms of separation for the internal design of switchboards. Larger electrical main distribution switchboards are usually specified as a type of Form 4 board as defined within BS 61439-2. This Standard provides varying degrees of safety from other terminals, switch fuses and the busbar when terminating another cable when the switchboard is electrically live. However, MCPs are commonly of a lower specification and may only be of the Form 2 type. This means that often terminals are accessible and there may be other live parts that are accessible too.

Consequently, whilst the maintenance activities on these control panels, and the motor sub-circuits, with respect to periodic testing, will be similar to the switchboards described in Section B1 of this Guide, extra safety precautions should be made to account for the live parts.

It is recommended that the reader refer to the isolation criteria of HSG 85 *Electricity at work: Safe working practices*, for example:

(a) does the MCP need to be live when testing is carried out on part of the electrical installation that supports the mechanical infrastructure?
(b) can it all be isolated to provide a safe, controlled environment for the maintenance team to work?

If the work has to be done whilst the MCP main isolator is switched on, then a full risk assessment must be done, the trained operatives must be fully briefed, appropriate PPE must be worn and correct insulated tools must be provided. It is also advisable for any operative to be accompanied by someone who has received first aid training especially with respect to electrical shocks.

B4.4 Machinery

The Provision and Use of Work Equipment Regulations 1998 (PUWER) requires that equipment provided for use at work is:

(a) suitable for the intended use.

(b) safe for use, maintained in a safe condition and inspected to ensure that it is correctly installed and does not subsequently deteriorate.

(c) used only by people who have received adequate information, instruction and training.

(d) accompanied by suitable health and safety measures, such as protective devices and controls. These will normally include emergency stop devices, adequate means of isolation from sources of energy, clearly visible markings and warning devices.

(e) used in accordance with specific requirements, for mobile work equipment and power presses.

With respect to the maintenance of machines and equipment, PUWER requires that:

(a) all work equipment be maintained in an efficient state, in efficient order and in good repair;

(b) where any machinery has a maintenance log, that log is kept up to date; and

(c) maintenance operations on work equipment can be carried out safely.

A risk assessment should be used to determine both the frequency and nature of maintenance. Account should be taken of the following factors:

(a) the manufacturer's recommendations;

(b) the intensity of use;

(c) operating environment (for example, the effect of temperature, corrosion, weathering);

(d) user knowledge and experience; and

(e) the risk to health and safety from any foreseeable failure or malfunction.

Additional information on this can be found at:

- http://www.hse.gov.uk/work-equipment-machinery/puwer.htm
- http://www.hse.gov.uk/work-equipment-machinery/maintenance.htm

The European Machinery Directive 2006/42/EC applies to all machines and requires that machines:

(a) undergo conformity assessment as required by the Directive;

(b) meets all of the relevant essential health and safety requirements for the product;

(c) be CE marked;

(d) are accompanied by a declaration of conformity and user instructions in the language of the end user; and

(e) that a technical file is compiled to demonstrate compliance with the above processes and requirements.

It should be noted that the Directive could apply if the maintainer alters the function or performance of a machine. However, if the function or performance has not changed, this is classed as a repair and no action other than a risk assessment and compliance with PUWER is required. So, for instance, changing a machine in a complex assembly or a component in a single machine as part of a maintenance exercise would not fall within the remit of the Directive.

BS EN ISO 12100:2010 *Safety of machinery. General principles for design. Risk assessment and risk reduction* provides the tools necessary to design and develop reliable equipment that remains fit for purpose throughout its lifecycle.

It outlines general principles of machinery safety and also the management of risk assessment. A three-step framework is provided to identify and eliminate hazards at different stages of the machinery's lifecycle.

Step 1: elimination of the hazard on the basis of design measures (inherently safe design measures).
Step 2: reduce risk through application of technical and complementary protective measures.
Step 3: warn against residual risks.

The following link from the Health and Safety Executive will provide more information on standards relating to machinery: http://www.hse.gov.uk/electricity/standards.htm

B4.5 Automation, control and instrumentation

Systems that provide automation, control and instrumentation are usually bespoke adaptations based on the manufacturer's proprietary systems. Maintenance of these systems is very dependent on the instructions and procedures left with the O&M manual for a particular installation.

It is important that any software, firmware or hardware have strategies to ensure continuing compatibility and support. The IT Infrastructure Library (ITIL) is essentially a series of documents that are used to aid the implementation of a framework for IT Service Management. ITIL v3 2011 provides a framework for IT Continuity Management and Change Management. More information can be found at http://www.itil.org.uk/.

Technology develops quickly and can be quickly superseded. Locking suppliers into long-term contracts relating to maintenance and overhaul automation, control and instrumentation systems may be seen as unreasonable.

Likewise, it may not be economic to specify closed protocol systems that lock an installation into one manufacturer and a small list of 'approved' maintainers. Whilst no system is truly open protocol, some systems do come close to this and at least provide for a larger list of qualified suppliers who can provide competitive maintenance contracts and innovative solutions as the installation evolves.

B4.6 BMS

Increasingly, building management systems (BMS) are being used not just to provide initial first line control, switching and system status feedback loops, but also to monitor energy usage, target lower consumption, and provide maintenance response to systems that are not following typical patterns. Such infrastructure is referred to as BEMS – Building Energy Management System.

There are various high level technical considerations to ensure that the BEMS provide all the necessary features and is also procured at the optimum cost. Notwithstanding any government demands to reduce CO_2 emissions, continually rising prices of energy and the increasing demands of newer, more advanced equipment make it more imperative than ever that the building operators find ways to more efficiently use, monitor and reduce its use of energy.

Particular standards in this area include:

- BS EN 15232:2012 *Energy performance of buildings. Impact of Building Automation, Controls and Building Management*
- ISO 16484-5:2007 *Building automation and control systems - Part 5: Data communication protocol*
- BS EN 15217:2007 *Energy performance of buildings. Methods for expressing energy performance and for energy certification of buildings.*
- ISO 13790:2008 *Energy performance of buildings - Calculation of energy use for space heating and cooling.*
- ISO 16484 *Building automation and control systems (BACS) – series of seven standards relating to function, implementation, hardware and other design and operational issues*
- BS EN 15603:2008 *Energy performance of buildings. Overall energy use and definition of energy ratings.*
- BS EN 15500:2008 *Control for heating, ventilating and air-conditioning applications. Electronic individual zone control equipment*

BS EN 15232:2012 in particular provides "a structured list of Building Automation and Control System (BACS) and Technical Building Management (TBM) functions which have an impact on the energy performance of buildings". Systems can then be assessed as class A to D inclusive. It is likely that most legacy systems will be class D, whilst new, fully integrated, state-of-the-art systems will be class A.

O&M manuals should provide instructions on:

(a) equipment safety checks;
(b) start-up and close-down procedures;
(c) daily operation; and
(d) full descriptions of operating features.

Manuals should also give full descriptive and maintenance details on each and every item of equipment supplied.

Increasingly, suppliers of proprietary BMS equipment have become aware of the need to incorporate legacy equipment as new systems are developed. This reduces costs to end users and means that wholesale replacement of sensors and controllers and other equipment is not required.

Developments to BMS technology and design philosophy continue and the integration of numerous building services is becoming a more viable method of improving both energy efficiencies and cost efficiencies.

Using a BEMS system allows the end user to set automatic monitoring and targets (M&T) parameters to drive the energy efficiency agenda. So, by monitoring through sub-metering in a particular area, targets can be set to drive down energy usage. BEMS may also monitor actual running hours of a particular plant so that the preventative maintenance programme can be run more efficiently.

Lighting control systems can be linked into BEMS, together with the access control system to reduce the number of luminaires still on in an unoccupied area of the building. This could be especially useful in multi-tenanted buildings, with restricted access arrangements.

Maintenance of such systems is specialised. Unless in-house members of staff have received accredited training directly from the BEMS manufacturer, it is suggested that outsourcing this activity may be desirable to either the manufacturer's own team (typically closed protocol systems) or to a systems house (typically open protocol).

Various items need to be periodically checked including:

(a) electrical panel maintenance – includes visual inspection of the panel installation and appropriate testing and inspection;
(b) hardware maintenance – replacement of components where necessary;
(c) network checks – connectivity of BMS infrastructure and capacity;
(d) software maintenance – updates to control strategies where new equipment has been added and software updates and back-ups where required;
(e) building performance maintenance – assessment of results, trends and patterns of energy use; and
(f) system integration with other services – for instance isolation of ventilation systems and other plant on activation of the fire alarm.

APPENDIX C

Summary of relevant legislation and standards

1 Health and Safety at Work Act 1974
2 Electricity at Work Regulations 1989 (Statutory Instrument No 635)
3 HSR25: *Memorandum of guidance on the Electricity at Work Regulations 1989*
4 HSG 85: *Electricity at work: Safe working practices*
5 Regulatory Reform (Fire Safety) Order 2005
6 The Construction (Design and Management) Regulations 2015
7 Health and Safety (Safety Signs and Signals) Regulations 1996
8 European Council Directive 92/58/EEC
9 BS 5499-10:2014 *Guidance for the selection and use of safety signs and fire safety notices*
10 BS 7671:2008+A3:2015 *Requirements for Electrical Installations*
11 BS 5839-1:2013 *Fire detection and fire alarm systems for buildings. Code of practice for design, installation, commissioning and maintenance of systems in non-domestic premises*
12 BIP 2109:2013 *The design, installation, commissioning and maintenance of fire detection and fire alarm systems in non-domestic premises. A guide to BS 5839-1:2013*
13 BS 5839-6:2013 *Fire detection and fire alarm systems for buildings. Code of practice for the design, installation, commissioning and maintenance of fire detection and fire alarm systems in domestic premises*
14 BIP 2044:2013 *The Design, Installation, Commissioning and Maintenance of Fire Detection and Fire Alarm Systems in Domestic Premises. A Guide to BS 5839-6:2013*
15 BS 5266-1:2011 *Emergency lighting. Code of practice for the emergency escape lighting of premises*
16 BS 6423:2014 *Code of Practice for the maintenance of Low Voltage Switchgear*
17 BS 8544:2013 *Guide for life cycle costing of maintenance during the in use phases of buildings*
18 BS EN 62402:2007 *Obsolescence management. Application guide*
19 BS EN 60445:2010 *Basic and safety principles for man-machine interface, marking and identification. Identification of equipment terminals, conductor terminations and conductors*
20 BS 6423:2014 *Code of Practice for the maintenance of Low Voltage Switchgear*
21 Environmental Protection Act 1990 Section 34 (Waste Management – the Duty of Care – A Code of Practice)
22 The Waste Electrical and Electronic Equipment (WEEE) Regulations 2013
23 BS ISO/IEC 27001:2013 *Information technology. Security techniques. Information security management systems. Requirements.*
24 BS 8580:2010 *Water quality. Risk assessments for Legionella control. Code of practice*
25 BS EN 50122-1:2011+A1:2011 *Railway applications. Fixed installations. Electrical safety, earthing and the return circuit. Protective provisions against electric shock*
26 BS EN 50308:2004 *Wind turbines. Protective measures. Requirements for design, operation and maintenance*
27 BS EN 82079-1:2012 *Preparation of instructions for use. Structuring, content and presentation. General principles and detailed requirement*
28 BS EN 61082-1:2015 *Preparation of documents used in electrotechnology. Rules*

APPENDIX D

References

1   Building Information Modelling (URN 12/1327)
    A UK government paper

2   www.bimtaskgroup.org

3   PAS 1192-3:2014
    Specification for information management for the operational phase of assets using building information modelling (BIM)

4   PAS 1192-3:2014
    Specification for information management for the operational phase of assets using building information modelling (BIM)

5   www.softlandings.org.uk

6   BG01/2007
    Handover, O&M Manuals, and Project Feedback - A Toolkit for Designers and Contractors

7   http://www.designingbuildings.co.uk
    (use search term 'Building owner's manual - O&M manual')

8   ITIL ® v3 2011
    (provides a framework for IT Continuity Management and Change Management)

9   IET Guidance Note 3: *Inspection and Testing*

10 IET Guidance Note 7: *Special Locations*

11 IET *Code of Practice for Electrical Safety Management*

12 IET *Code of Practice for Cyber Security in the Built Environment*

13 HTM 06 02 *Electrical safety guidance for low voltage systems*

14 HTM 06 03 *Electrical safety guidance for high voltage systems*

15 JSP 375 *Management of health and safety in defence*
    https://www.gov.uk/government/collections/jsp-375-health-and-safety-handbook

16 The Health and Safety (Safety Signs and Signals) Guidance on Regulations Second edition 2009 free download at http://www.hse.gov.uk/pubns/books/l64.htm

17 Scope of equipment covered by the UK WEEE Regulations (LIT7876) free download at https://www.gov.uk/government/publications/eee-producers-how-to-accurately-identify-eee

18 Health and Safety executive website information on Legionella
    http://www.hse.gov.uk/legionnaires/

19 HSE L8 ACOP 4th Edition 2013
    Legionnaires' disease: The control of legionella bacteria in water systems

20 HSE HSG 274
    Legionnaires' disease; Technical guidance

21 CIBSE Technical Memoranda
    TM13 - *Minimising the Risk of Legionnaires' Disease 2013*

22 Health Technical Memorandum 04-01:
The control of Legionella, hygiene, "safe" hot water, cold water and drinking water systems

Part A: Design, installation and testing

Part B: Operational management

23 BSRIA Legionnaires' Disease – Risk Assessment (BG 57/2015)

24 BSRIA Legionnaire's Disease – Operation and Maintenance Log Book (BG 58/2015)

25 ENA Engineering Recommendation S34.
A Guide for Assessing the Rise of Earth Potential at Substation Sites. 1986.

26 ENA Engineering Recommendation S36. Procedure to Identify and Record "HOT" Substations. 1994 and amendment 2007.

27 ENA Technical Specification 41–24.
Guidelines for the design, Installation, Testing and Maintenance of Main Earthing Systems in Substations. 1992.

28 CIBSE Applications Manual no.12 (AM12)
Combined heat and power for buildings.

29 IEC 479-1
Effects of Current on Human Beings and Livestock, IEC, 1994.

30 BEAMA
Guide to Surge Protection Devices (SPDs): selection, application and theory (2014)

31 HSE INDG 270 Supplying new machinery
A short guide to the law and your responsibilities when supplying machinery for use at work www.hse.gov.uk/pubns/indg270.pdf

Glossary

Term	Abbreviation	Comments
Authorised person	AP	Senior engineer responsible for safe systems of work on site; issues permits to work and monitors maintenance work.
Authorising engineer	AE	Audits safe systems of work, maintenance records, logbooks, assesses APs and oversees all maintenance works on site.
Building information modelling	BIM	Digital platform for multi-disciplinary design co-ordination, collaboration and subsequent collation of construction, commissioning and operational information in buildings.
Building management systems	BMS	An automated system of sensors and controls to monitor the status of building systems. Increasingly used to facilitate switching and energy monitoring too.
Building research establishment	BRE	Independent research, testing and training organisation with expertise in the built environment and associated industries.
Capital expenditure	CapEx	Funds identified by a business case and specifically set aside to redesign and refurbish an area of a building or estate, or a specific system such as access control or lighting.
Chartered Institution of Building Services Engineers	CIBSE	Professional engineering institution serving mechanical and electrical engineering design professionals and technicians.
Construction (Design and Management) Regulations	CDM R	CDM R is a framework for managing health and safety aspects of construction sites; it defines the roles and responsibilities for a number of stakeholders within the construction process.
Competent person	CP	Experienced technician responsible for carrying out maintenance work within the framework of a safe system of work.
Cyber security	–	Philosophy of protecting digitally connected systems from compromise and disruption by non-authorised persons.
Duty holder	DH	Person ultimately responsible for health and safety within a building or wider estate and for overseeing the management of the maintenance systems and procedures.
Hot water and cold water sentinel outlets	–	Sentinel taps are designated outlets at which water temperature testing is carried out on a regular basis – this can also be constantly monitored via the BMS.

Term	Abbreviation	Comments
Internet of Things	IoT	The Internet of Things is the provision of unique identifiers for objects, animals or people and the network capability to transfer data without requiring human-to-human or human-to-computer interaction.
Mean time before failure	MTBF	The manufacturer's estimate of the expected average time before failure based on particular environmental and operational conditions.
Operation and maintenance manuals	O&M	Documentation compiled of drawings, commissioning data, design specification and the manufacturer's data that acts as a point of reference for the operational team.
Operational expenditure	OpEx	Funds allocated for regular maintenance and for breakdown repairs.
Permit to work	PTW	A management process to regulate what work is carried out, isolations and reinstatements and to ensure a safe system of work is implemented.
Personal protective equipment	PPE	Equipment used to provide a minimum level of safety from injury for operatives such as hard hats, protective boots, gloves, goggles and ear defenders.
Power over ethernet	PoE	The use of ethernet cabling to provide electrical devices with a power connection instead of more conventional electrical power circuitry.
Predictive maintenance and monitoring		A process by which the performance of a system is carefully observed and maintenance is carried out when set parameters reach predefined levels.
Preventative maintenance		Replaces key components according to a schedule before they completely fail so that the whole system maintains an output that is close to the original design.
Reactive maintenance		Waits for key components of a system to fail before maintenance activities take place.
Risk assessments and also method statements	RAMS	Collectively known as RAMS, these two complimentary documents assess the dangers to life and property within a maintenance process, and provide control measures to mitigate or remove the dangers.
Safe systems of work	SSoW	A safe system of work is a collection of procedures designed to ensure the safety of maintenance operatives and those in the vicinity of maintenance activity.
Standard operation procedures	SOP	A predefined series of activities usually set out in chronological order for particular scenarios.

Term	Abbreviation	Comments
Toolbox talks		Describes the use of on-site briefings to reinforce previous training and experience or to introduce changes.
Value engineering	VE	The real definition of value engineering should be to achieve the same specified outputs at a lesser cost without reducing quality.
The rule of 9s		Data centres categorise their availability according to the rule of 9s (90 %, 99 % 99.9 %, 99.99 % etc.) measured over a year.

INDEX

Note: alphabetic letters refer to Appendices

access to plant	2.1.4; 2.3
as-built drawings	2.6; 5.3
assessment	5.2.1; 5.3
asset management	4.2.4
audits	
after maintenance activity	3.5
compliance checks	2.4
life safety systems testing	2.5.2
maintenance procedures	5.5
protective devices	B2.1
authorised persons (AP)	4.3.4; 5.5
authorising engineers (AE)	5.5
automatic monitoring: *see* computer-aided maintenance monitoring	
backlog lists	3.5; 5.4
BACS (building automation and control systems)	B4.5
BEMS (building energy management systems)	B4.6
benefits	3.2; 3.1
BIM (building information modelling)	2.2
BMS (building management systems)	4.2.3; B4.6
bonding connections	B1.7
BREEAM	1.5
British Standards	
BS 5266	2.5.2; B3.1
BS 5446-3:2015	B3.3
BS 5499-10:2014	4.3
BS 5839:2011/13	2.5.2; B3.2; B3.4
BS 6423:2014	4.2; B1; B1.1
BS 6626:2010	B1.1
BS 6651:1999	B1.8
BS 7671:2008/15	1.3; 3.4; 4.2.1; B1.7; B1.9; B2; B2.3
BS 7967:2015	B3.3
BS 8300:2009/2010	B3.4
BS 8544:2013	2.5
BS 8580:2010	B2.4
BS EN 13015:2001/08	B3.4

BS EN 15232:2012	B4.5
BS EN 50291-1:2010/12	B3.3
BS EN 50292:2013	B3.3
BS EN 50545-1:2011	B3.3
BS EN 60079:2007/14	B3.3; B4.1
BS EN 60422:2013	B1.2
BS EN 60445	2.4.1
BS EN 61082-1	2.4.1
BS EN 61439-2:2011	B4.3
BS EN 62305:2011	B1; B1.8; B1.9
BS EN 62402:2007	2.5
BS EN 81-28:2003	B3.4
BS EN 82079-1:2012	2.3.1
BS EN ISO 12100:2010	B4.4
BS EN ISO 19011:2011	5.5
BS EN ISO 82079-1	2.4.1
BS ISO/IEC 27001:2013	5.3.1
building automation and control systems (BACS)	B4.6
building energy management systems (BEMS)	B4.6
building information modelling (BIM)	2.2
building management systems (BMS)	4.2.3; B4.6
building performance	1.5
Building Regulations Part L	1.5
business impacts	5.5
bypass supplies	3.4
call systems	B3.4
capital expenditure (CapEx)	2.1.6
carbon monoxide detectors	B3.3
CDM (Construction (Design and Management)) Regulations 2015	1.2; 2.3.1; B
certification	5.2
CHP (combined heat and power)	B1.5
circuit charts	2.4.1; 2.2
circuit protective conductors	B1.7
circuit protective devices	B2.1; B2.1
combined heat and power (CHP)	B1.5
commissioning	2.3; 2.6; 5.3
compatibility of replacements	2.5
competent persons (CP)	4.3.4; 5.5
compliance checks	2.4
component compatibility	2.5

component deterioration | 3.1; 3.1; 4.2.3; 5.3

component obsolescence | 2.5; 3.2

component preferences and availability | 2.1

component selection | 2.1.6; 2.5; 4.2.2; A1.4

computer-aided maintenance monitoring | 4.2.4; 5.3.1
    see also information technology (IT)

Construction (Design and Management) (CDM) Regulations
    2015 | 1.2; 2.3.1; B

contingency planning | 3.3

contractors | 2.1.5; 3.3

control equipment | B4

control of maintenance | 3.2; 3.4

control panels | B4.3

corrective maintenance | 4.2

correlation (identities) | 2.4

costs | 2.1.6; 2.5; 5.6

cyber security | 4.2.4; 5.3.1

DC systems | B2.5

dead working | 4.3.4; 4.1
    see also electrical isolation

derogations | 2.1

design drawings | 2.3.1; 2.6

design for maintenance | 2.1; 2.1; A1

diagnostic fault trees | 2.4.2

disabled toilet alarms | B3.4

disconnection of supply: *see* electrical isolation; service
      interruptions

documentation

    commissioning | 2.6

    electronic document storage | 2.3.1

    maintenance activity | 2.5; 3.5

    secure access to | 2.4.2; 5.3.2

    domestic hot water | B2.4; B2.5

    downtime | 2.1; 3.1
    see also drawings; logbooks; operation and maintenance
      (O&M) manuals

drawings | 2.3.1; 2.4; 2.6; 5.4
    see also circuit charts; documentation

earthing conductors   B1.7; B1.5

economic considerations   2.1.6; 2.5; 5.6

electrical isolation   3.4; 4.3.4

electrical maintenance, definition   1.6

Electricity at Work Regulations (EWR) 1989   1.2; 3.4; 4.3

electronic document storage   2.3.1

emergency lighting   1.2; B3.1; B3.1

emergency voice communication systems   B3.4

energy performance   1.5

Environmental Protection Act 1990   4.3.8

equipment failure modes   5.2.3; 5.5

equipment preferences   2.1
   see also component selection

equipment resources   2.1.5

European Machinery Directive 2006   B4.4

evaluation of electrical systems   5.2

EWR (Electricity at Work Regulations) 1989   1.2; 3.4; 4.3

failure modes evaluation   5.2.3; 5.5

failure scenarios   5.2.4; 5.3

fault correction: see remedial activities

fault protection   B2.1; B2.1

fault trees   2.4.2

fire detection and alarm systems   1.2; B3.2; B3.3
   see also life safety systems

fire suppression systems   B1.1

first line response   2.4

follow-up checks   3.5

functional tests   5.2; 5.2

gas/carbon monoxide detectors   B3.3

general lighting   B2.2; B2.2

generators   3.4; B1.3; B1.2

handover   2.2; 2.6

harmonic filters   B1.11; B1.7

hazardous areas and equipment   B4.1

Healthcare Technical Memorandum (HTM)   B3.4

Health and Safety (Safety Signs and Signals) Regulations 1996   4.3

Health and Safety at Work etc. Act (HASAWA)1974   1.2

health and safety risks   4.3.1

heat pumps	B2.4
high voltage: *see* HV	
hot water heating	B2.4; 2.5
hot water services	B2.4; B2.5
housekeeping	3.5; 4.3.8
HSG 85	4.3
HSR 25	2.1.4; 4.3
HV installations	B1.1
HV/LV transformers	B1.2
identification systems	2.4; 2.6; 2.2
implementation	5.2.4; 5.6
industrial equipment	B4
information technology (IT)	2.5; B4.5
see also computer-aided maintenance monitoring	
in-house maintenance staff	2.1.5; 2.4
installation schedules	5.2
interlocks	2.4.2; 4.3.4
Internet of Things (IoT)	4.2.3; 5.3.1
isolation, electrical: *see* electrical isolation	
isolation transformers	B1.2
key operated interlocks	2.4.2; 4.3.4
labelling systems	2.4; 2.6; 2.2
examples	2.8
sample schedule	2.7
see also safety signage	
layout drawings	5.4
legacy systems	2.5; 4.2.2
legislation	1.2; C
lifecycle costing	2.5
lifecycle design	2.2
life safety systems	1.2; 2.5.2; 3.1; 5.5; B3
lift communications systems	B3.4
lighting	B2.2
lightning protection systems	B1.8
live working	3.4; 4.3; 4.1
local standards	2.1
lock-offs: *see* safety locks	
logbooks	2.3.1; 2.5.1; 3.3

luminaires   B2.2; B3.1

machinery   B4.4
   see also generators
mains earth terminals (MET)   B1.7
main switchgear and switchboards   B1.1; B1.1
maintainability   1.3
maintenance activities   3.4; 3.1; A2.2; A2.3
   audits   5.5
   before the event   A2.1
   during the event   A2.2
   following the event   A2.3
   frequency   1.3; 2.1; 5.2; Appendix B
   general applications and circuit protection   B2
   industrial and control equipment   B4
   life safety systems   B3
   supply intake   B1
   see also management of maintenance, planning of
     maintenance, remedial activities
maintenance backlogs   3.5; 5.4
maintenance factors (lighting)   B2.2
maintenance records: *see* documentation
maintenance resources: *see* resource planning
maintenance safety: *see* safe systems of work (SSoW)
maintenance staff: *see* personnel
maintenance strategies   4.2; 5.2; 5.6
management of maintenance   3.2; 3.4
manual handling   2.1.4
manufacturer's information   2.3; 5.2.2; 5.4
manufacturers' instructions   2.3.1; 5.4
manufacturers' warranties   3.1; 3.1
market availability   2.1
mean time before failure (MTBF)   2.5
mechanical interlocks   2.4.2
MET (mains earth terminals)   B1.7
method statements   2.4.2; 3.3; 4.3.1; 4.3.3; 4.3.6
micro generation   B1.6; B1.4
monitoring systems
   computer-aided   4.2.4; 5.3.1
   manual or automatic   2.4

for predictive maintenance	4.2.3
remedial activities	5.4
motor control panels (MCP)	B4.3
motors	B4.2
MTBF (mean time before failure)	2.5
multiple supply intakes	2.4.1; 2.4.2; B1
notifications	3.3
O&M: *see* operation and maintenance (O&M) manuals	
obsolescence	2.5; 3.2
occupant notifications	3.3
operational access	2.1.4
operational expenditure (OpEx)	2.1.6
operational performance	5.3
operational restrictions	2.1
operational risks	4.3.2; 4.1
operational security	2.4.2; 5.3
see also cyber security	
operation and maintenance (O&M) manuals	2.3.1; 3.2
commissioning data	5.3
content	2.3.2; 2.6; B; B4.6
manufacturer's information	5.2.2
outsourcing	2.1.5; 3.3
overload protection	B2.1; B2.1
patient call/nurse call systems	B3.4
performance degradation	3.1; 3.1; 4.2.3; 5.3
performance gap	1.5
periodic inspection and testing	4.2.1; 5.2; 5.1
permits to work (PTW)	2.4; 3.3; 3.5; 4.3.5
personnel	
briefing	3.4; 4.3.3
debriefing	3.5
resources	2.1.5; 5.3
training and familiarisation	2.4; 3.3
see also resource planning	
personnel alarms	B3.4
PFC (power factor correction)	B1.10; B1.6
pictorial diagrams	2.4.2

| planning of maintenance | 3.4; 5.2.4; 5.6; A2.1 |
| | |

see also resource planning

plant access	2.1.4
PoE (power over ethernet)	4.2.3
post-fault maintenance	4.2
power factor correction (PFC)	B1.10; B1.6
power over ethernet (PoE)	4.2.3
power supplies	B1
predictive maintenance	4.2; 4.2.3
preventative maintenance	4.2; 4.2.1
product roadmap (system upgrades)	2.5
programme planning	5.6
Provision and Use of Work Equipment Regulations 1998 (PUWER)	B4.4
PTW (permits to work)	2.4; 3.3; 3.5; 3.5

| RCD (residual current devices) | B2.3 |
| reactive maintenance | 4.2; 4.2.2 |

reconnection: see system reinstatement

regulations: see Building Regulations; see legislation; see Wiring Regulations

Regulatory Reform (Fire Safety) Order 2005	1.2; 2.5.2
remedial activities	3.4; 3.5; 4.2.2; 5.4
renewable energy	B1.6; B1.4
replacement parts	2.5; 4.2.2

see also spare parts

residual current devices (RCD)	B2.3
resource planning	2.1.5; 2.4; 2.1; 3.2; 3.3; 5.3; A1.3

see also personnel

| responsibilities | 1.2 |

reviews: see audits

risk assessments	2.1.3
standard operating procedures (SOP)	2.4.2
risk assessments	2.1.3; 3.3; 4.2.2
derogations	2.1
earthing and bonding connections	B1.7
example	2.2
failure modes evaluation	5.5
health and safety	4.3.1

operational   4.3.2; 4.1

    see also method statements

risk registers   3.5; 5.4

rule of 9s   2.1

safe access and egress   2.1.4; 2.3

safe systems of work (SSoW)   2.4; 3.3; 4.3

safety impacts   5.5

safety locks   2.4.2; 4.3.4

safety signage   4.3

scenario planning   5.2.4; 5.3

schematic drawings   5.4

security issues   2.4.2; 5.3.2

    see also cyber security

service interruptions   3.2; 3.3; 4.3.2; 4.3.6; 4.3.7

short term operating reserve (STOR)   B1.5

signage   4.3

site-specific information   2.6; 5.4

small power   B2.3; B2.3

Soft Landings   2.2

solar PV   B1.6; B1.4

SOP (standard operating procedures)   2.4.2; 2.6; 5.3.2; 5.6

space planning   2.1.4; 2.3; A1.2

spare parts   4.2.2; 5.6

SPD (surge protection devices)   B1.9

specialist contractors   2.1.5

specialist equipment   2.4

specialist training   2.1

SSoW (safe systems of work)   2.4; 3.3; 4.3

stakeholders   2.1.2; A1.1

    consultation   2.5; 3.3; 5.2.4

standard operating procedures (SOP)   2.4.2; 2.6; 5.3.2; 5.6

standards   1.5; 2.1

    see also British Standards

standby supplies   3.4; B1.3; B1.2

statutory requirements: *see* legislation

STOR (short term operating reserve)   B1.5

structural loads   2.1.4

supervision   3.4; 4.3.4

supply changeovers	2.4.2
supply intake	B1
see also multiple supply intakes	
surge protection devices (SPD)	B1.9
surveys	5.3
sustainability	1.5
switchgear and switchboards	B1.1; B1.1
switching schedules	2.4.2
switchrooms	B1.1
system degradation	3.1
system design	2.1
system reinstatement	3.2; 3.4; 3.5; 4.3.6; 4.3.7
system status assessment	5.2.1; 5.3
task briefing and management	3.4; 4.3.3
technical building management (TBM)	B4.6
temporary supplies	3.4
testing	5.2
types and frequency	5.1; 5.2
see also periodic inspection and testing	
training	2.1; 2.4; 2.6; 3.3
transformers	B1.2
uninterruptible power supplies (UPS)	2.4.2; B1.4; B1.3
user confidence	3.1
user consultation	3.3
user reassurance	3.5; 4.3.7
value engineering (VE)	2.5
variable speed drives (VSD)	B4.2
verification tests	5.2; 5.1
waste disposal	3.5; 4.3.8
Waste Electrical and Electronic Equipment (WEEE) Regulations	
2013	4.3.8
Wiring Regulations	1.3
workflow diagrams	2.4.2
working methods: *see* method statements; *see* safe systems of	
work (SSoW); *see* standard operating procedures (SOP)	